OPERATIONS ANALYSIS OF ENGINEERING SCIENCE

MISSION OF LAWRENCE LIVERMORE NATIONAL LABORATORY

BY

Dr. Ronald W. Cutburth

Dissertation Submitted in Partial Fulfillment of

The Requirements of the Degree of

Doctor of Philosophy

Administration and Management of Engineering Science Operations

Walden University

ABSTRACT

This is an analysis of the mission of the Lawrence Livermore National Laboratory

(LLNL). The study examined the Engineering Sciences Operation's technical skill,

operations methods, functions, and needs required to achieve the laboratory mission.

A survey of the main programs and projects was undertaken in order to understand the

attributes of an ideal research agency, toward the future benefit of the United States.

A unique aggregate of research and innovation production methodology was discovered.

This is examined using a new model developed by the author and titled Concept Fusion.

The model demonstrates the one-to-one correspondence of a knowledge explosion

potential, and its relation to the principles of nuclear fusion. A relationship of theoretical

mathematical principles, in the form of a pure mathematical linguistic, is presented for a

detailed understanding of the underlying connections of knowledge growth, nuclear

theory, and mathematically based science concepts. Optimization and expansion of the

current research at an ideal agency is suggested. An innovative science is suggested,

through adaptation of the interconnected principles of theoretical mathematics, language

and the sciences. A meta-analysis was used to examine the LLNL mission. An aggregate

of analysis categories is collected to provide triangular validation for each of the

categories. Two primary areas from the aggregate are analysis of the engineering sciences

operations using operations models, and collection and consideration of the legal and

political forces that affect the operations. The Concept Fusion model provides a third

perspective, for triangular validation, of the needs and potential of an ideal national

laboratory.

It is suggested that a paradox exists in the operations at LLNL. It was shown that, for the mission to be accomplished (which includes nuclear weapons stewardship), almost all science development is required, which in turn is usable for all technology development, that is actually not nuclear weapon-centered. The paradox resulted from the interconnectivity of the elements of knowledge and purely mathematical principles. These are generic to a focus, and offer expansions of knowledge in any category- known to humankind. This is delineated in the Concept Fusion model

ACKNOWLEDGMENTS

To:

Dr. Spiros Dimolitsas, Co-director of Lawrence Livermore National Laboratory (LLNL).

For his assistance and information duty delegations about the engineering sciences operations.

Dr. Robert Langland, Technical specialist, Lawrence Livermore National Laboratory.

For his assistance and information about the LLNL complex research operations.

The Lawrence Livermore National Laboratory staff and executive secretaries.

For filling the information and communications network needs. The needed information was expeditiously and professionally handled. I was made to feel welcome on all calls and visits.

Gary Boss, Assistant Director, General Accounting Office (GAO).

For his professional and personal instruction on GAO research and evaluation methods.

Dr. Gene Woolsey, Colorado School of Mines.

For his assistance and guidance as my mentor through this program until the dissertation was essentially complete. His experience in the combined engineering sciences and operations analysis proved to be fundamentally effective. His dedication toward my success exceeded the boundaries of those essential needs.

TABLE OF CONTENTS

LIST OF FIGURES

NOTICE

Many charts and graphs are removed to focus on the core comparison of the LLNL or US government knowledge growth model compared to Dr. Cutburth's. Concept Fusion model for growing Concepts.

The author opines and shows in his dissertation evidence of thoughts that Concepts are understood to be at a higher level of thinking than simply collecting data as knowledge in many cases.

After about 20 years since his publication of his dissertation. The US has accumulated massive debts with little going toward Research and Development. Defense spending and wars have depleted the US economy and allowed millions more to become unemployed and on welfare. The author opines the responsibility falls on Congress.

ASSUMPTIONS

Large questions were shown in public documents, with regard to the Lawrence Livermore National Laboratory mission. To encompass the question of the mission, this research program was deliberately broad. However, there were two assumptions made.

1. Although a prior study by a corporation leader concluded LLNL should function

   more like a business, there was information showing that la
   The analysis was based in part to compare the practices of LLNL given this

   limitation. It was found that LLNL management uses management "best

   practice" methods, but federal laws do limit the LLNL

   acting like a business.

2. Although the LLNL claimed mission is predominantly with regard to nuclear

   weapons, preliminary search of records showed that they may be involved in

   multiple categories of research that are not classified. It was assumed that a

   search would demonstrate useful non-classified research information, and this

   was proven to be true.

CHAPTER 1
INTRODUCTION TO THE STUDY

The United States funds and directs operation of national laboratory research and development facilities. The primary direction of these laboratories is given by federal agencies, primarily the Department of Energy (DOE). Also, an essential part of the research and development conducted is directed by the Department of Defense (DOD). Congress evaluates its funding of these federal agencies on a periodic or yearly basis. Congressional hearings are conducted which provide the collection of pertinent data and consensus, to aid in Congressional decision making toward funding choices. In addition, the hearings provide evidence for Congress that can aid in determination of need to create or revise laws, with regard to appropriate national good. Therefore, all of the pertinent decisions are made by Congress for the funding of the national laboratories.

Recent Congressional hearings have a documented concern that the laboratories are not managed well. The General Accounting Office (GAO) has recorded reports and claims to challenging the current management practices used by the laboratory directors and/or the DOE and DOD. Congress has responded with hearings and is considering whether the national laboratories may need to be restructured.

Simultaneous hearings have been conducted for discovery of challenges to the Department of Commerce and its divisions, including the National Institute of Standards and Technology (NIST). These agencies are involved in functional operations interrelated with national laboratory operational information sources. Reorganization of these agencies is also under consideration.

In every reorganization consideration, there seems to be a clear pattern: Congress is being presented with evidence that major difficulties may exist. In particular, significant differences exist in reports from studies about the analysis of the organizations. For example, the General Accounting Office (GAO) and the Office of Management and Budget (OMB) conducted an audit of the national DOE uranium.

enrichment operations. In one of the topics of focus, the GAO evaluated recovered dollars from the sale of enriched uranium. The GAO determined there was an unrecovered cost of $8.8 billion compared to $3.5 billion note by the OMB. The difference in accounting was due to the issue of whether the DOE could write off around $4.0 billion in costs of enriching uranium. (See the reports noted below, Alternative Futures for the Department of Energy .) But Lawrence Livermore National Laboratory (LLNL) developed a less expensive method, referred to as the Advanced Laser Isotope Separation (AVLIS) process. This process development resulted in a potential savings for isotope enrichment. However, the LLNL research was listed as cost factor, even though LLNL developed a less expensive process.

Significant differences in accounting practice warrant that the basic standards of accounting should be examined. Furthermore, one could examine the entire premise for the reasoning behind the accounting. This can be considered with reference to the functional purpose for having research laboratories. In other words, since fundamental accounting can be questioned, then consideration should be made for the philosophical basis for the items and ideas to be accounted for. The determination of the value of the research effects the laboratory missions.

For example: the national laboratories claim significant research, education, and technology development. This significant function produces results that ought to be evaluated. However, this has not been accurately measured in practice, and is one of the issues in findings from the Congressional hearings. The 1995 Congressional hearings and reports Federal Technology Transfer Policies, and

Federal Laboratories: Methods for Improving Incentives and Alternative Futures for the Department of Energy indicate that controversy simultaneously exists about management methods, national policy, and the mission of the national laboratories.

One questions the validity of the overall mission plans when comparing national accounting methods and testimony of other government agencies.

The July 1995 Congressional report Reshaping the Graduate Education of Scientists and Engineers: NAS's Committee on Science, Engineering, Public education Policy provides a separate assessment of education needs. The needs assessment of national education policy is also in question. Therefore, one finds no clear alignment with the research reports noted earlier, when there is controversy with regard to the basic missions. In addition, one can examine the Congressional report and hearings with regard to the National Institute of Standards and Technology (NIST) budget. This agency claims sponsorship of research through companies and universities. But there is controversy as to whether this agency should be disbanded.

As shown in the above reports, policy on research and the central missions of the national laboratories are controversial. In the Congressional hearings and reports, the LLNL director claimed that nuclear weapon stockpile maintenance is the central Mission However, LLNL has been governed by the DOE for more than 20 years, and the central mission of the DOE is energy, not nuclear weapons. Furthermore, statements in the hearings noted above infer that the LLNL central mission seems to have changed. In addition to questions about the LLNL mission, reports given in the above hearings claim observance of inefficiency at LLNL, but conclude that this is caused largely by the DOE and Congressional leadership. Furthermore, LLNL also claims significant benefit to industry and education in sciences. But the contributions are suggested to be broad, and not indicated to be directly related to nuclear weapons stewardship.

In other words, it appears that even though the central mission for LLNL is supposedly nuclear weapons maintenance, this claim should be examined, because the LLNL mission also involves national education benefit planning. This gives cause to consider the means of accounting, because the determination of value for a mission is not necessarily those for funds expenditures, and regular accounting practices. What, then, are value, mission, process, and operation?

It is argued that operations analysis of current sciences practices and missions require a broader basis, in order to overcome the controversy created by the value analysis found in prior accounting methods. A meta-analysis (providing triangular validation) of the operations must be broad, when the operations are shown to be broad and scientifically complex. The parameters must include integrate analysis of overall research and development operations, principles of knowledge and innovation production, and consideration of the political and legal forces. All of these affect the operations plans for the mission.

PROBLEM STATEMENT

1. A clear national laboratory mission statement has not been isolated, as shown by the Congressional hearing records and the April 1995 GAO report Department of Energy: National Laboratories Need Clearer Missions and Better Management. A clear mission statement is essential to discover an effective plan of operations, for determination and actualization of a mission. A conflict exists in the government reports when the LLNL claims it has a mission, and the GAO report states that the laboratories need clearer missions. A conflict of overall management methods and policy exists when the GAO report suggests that the laboratories should operate more like businesses, and that they are not managed well. This conflict becomes more acute when government regulations state that laboratories are not permitted to operate as businesses. All of the issues create the need to study the mission

2. A clear plan for science education policy and practice needs to be considered with regard to clarified or upgraded mission statement. The primary purpose of this project is an analysis to discover the mission.

BACKGROUND

Boundary extremes, which simultaneously define global proportions in applications and nuclear particles in the minutes measurement frame of reference, can be used to describe the background for functionally complex research operations. Complex variables, as a concept definition, can be used to perceive and develop an understanding of the interrelationships for advanced research and development operations, as shown in chapter 4 of this project.

In light of the boundary extremes needed for national research, Congress is examining the overall research programs and planning. The DOE, and the national laboratory R&D activities, have been challenged with regard to their efficiency and the GAO has generated a report about the DOE missions. and its laboratories. The report, Department of Energy: National Laboratories Need Clearer Missions and Better Management, was derived from GAO research that was ordered by Congress. Therefore, there is a Congressional record showing that the issues and controversy about the laboratory missions are significant.

The activities for research can vary in intensity, depending on the form of the operation or mission of the research facility. Clearly, nuclear research for both atomic energy and nuclear weapons can be considered as research at all boundary extremes. The LLNL annual report for Congress, Institutional Plan, FY 1996-2001, demonstrates immense breadth in research categories. For example: consideration of the nuclear interaction of material sciences is at the basis, and minute extreme, of the materials and processes found in life. Simultaneous research for absolute global war and ecodynamics is at the widest extreme. Both extremes can be simultaneously examined in a complex research plan. This is demonstrated in the research and analysis of the operations that exist at LLNL, and is analyzed and summarized in Chapter 4.

Operations needs and analysis of broad research activities produce interactive, complex variables for management considerations. These are found in the national laboratory research environment of LLNL. The University of California's triad of Lawrence Livermore, Lawrence Berkeley, and Los Alamos national laboratory group defines a broad and complex system. A system of four laboratories, including the Livermore Laboratory neighbor—Sandia National Laboratory—is the more accurate quantum of science activity. (Sandia is a neighbor of LLNL in California, and has operations in New Mexico).

Regional planning, science education, engineering sciences operations, and fundamental research activities have interactive parameters and needs for the breadth of research in this form of laboratory environment. In addition, technology advancement for the economy and technological are active components of the LLNL systemic operations, as shown in the data comparisons that are reported in Chapter 4.

Given all of the elements for the laboratory operations noted above, laws and legal systems become an integral part of the regional operations planning and developments. The system of combined interactivity for governing and operational law parameters impact the laboratory complex. Clearly, the collection of all of the noted parameters are driven by societal needs, which have accelerated as rapidly as the sciences have advanced from laboratory research.

In other words, research has produced both advanced and complex technical developments, as well as advanced societal awareness of the advantages of these developments (with some implied disadvantages). The net result for science development of the computer age, the space age, and the atomic age create a collected information age.

One could say that society is now awake, and asks about all the complexities derived from fundamental, Interactive, and broad research and development. This study is not of history, it is of the current operational requirements and activities for responsible and effective research and development.

PURPOSE

1. To evaluate the engineering sciences operations to determine the mission of national laboratories, focusing on LLNL, including analysis of the mission statement. Current activities that may clarify or eliminate the conflict found in government publications about the LLNL mission and its management methods were examined examined.

2. To evaluate the engineering sciences research and development operations, for determination of its current activities and capabilities of supporting the LLNL mission. The possibility of an upgraded mission statement for LLNL will be considered. This was an evaluation of the simultaneous existence of the mission, or a resultant actual mission.

A list of general research questions that are created from the above research questions are shown next. These are stated here and in Chapter 2 following the methodology statements. The general questions answered with this research follow.

SYSTEMS AND OPERATIONS QUESTIONS

Regarding the Engineering Sciences Operations Policy and Planning

A list of general questions about the LLNL operations and mission follows. These are designed to focus on significant issues that are examined in this research project and are answered in Chapter 5. The general questions about the LLNL operations are to help determine a placement of LLNL in the national community of activities.

1. Do the operations follow norms found in nongovernment entities (industry)? Can it be said that the operations parallel business operations?

2, Do the operations experience catastrophic developments due to changes in government policy? Or, does the operations design allow proportionate changes? Can a change impact be compared to market changes for industry? Is LLNL able to respond to changes effectively?

3. Does the military project research planning allow technology adaptation for non-military use? Does the operation plan lend toward a clear dual use benefit?

4. Does the State of California provide financial support of the LLNL research operations?

5. Are systemic needs internally consistent—toward efficiency and creativity?

6. Do information sciences technologies and planning permeate the operations?

7. Given a known nuclear weapons responsibility, can one find a central function which is not actually nuclear weapons oriented?

8. Do technology skill areas seem to operate at the same quality level in all activity areas?

9. Does the LLNL location appear to be appropriate? Can significant location questions be found?

10. Do the operations provide benefit to national education?

11. Can technology research areas be found which may be effectively divested or added?

Answers to the above questions are expected to be general. Some questions may produce a set of added questions that could be expected to provide useful information for futures planning and research

SIGNIFICANCE

National energy needs, a nuclear weapon stockpile, and fundamental product development research are considered as reasonable topics for evaluation and restructuring. The reports noted above show that all of the U.S. funded research facilities are in review. In particular, essential elements of the national research missions are in review. Simultaneous restructuring, and potential dismantling, are in consideration for all agencies that sponsor and control fundamental research. The actual results and timing of restructuring, or whether it occurs, will be best served by significant amounts of analysis material about laboratory missions to support or oppose any Congressional action.

♦ This project was designed to assist Lawrence Livermore National Laboratory (LLNL), by analysis of significant data and analysis tools, to carry out its mission and engineering sciences research operations.

♦ This project was designed as a strategic operations and management analysis program. It is developed from materials provided for this researcher by the Lawrence Livermore National Laboratory Engineering Sciences Operations Director (LLNL-ESOD).

♦ This project was designed to assist congress in decisions about the LLNL mission.

♦ This project was designed to critique current management methods with regard to potential application in analysis of a national laboratory.

♦ This project was designed to discover the basis of an ideal research laboratory, for adaptation to future national research planning

NATURE OF STUDY

This project was designed to be a meta-analysis. It uses existing non-classified reports and summary material to provide significant information for the study. This material addresses the overlap of interacting forces that affect the engineering sciences operations, the mission interpretation, and LLNL actualization methods. Significant research project evidence was found and used in a cross-correlation for new determinations about the effects of planning and methodology on knowledge production and subsequent innovation production. Facts found in Congressional hearings, the national code, and engineering sciences field

Many texts were used to develop a new understanding of the interacting needs for examining the mission definition and its subsequent actualization.

SCOPE AND LIMITATIONS

This project includes evidence available from published sources about the programs and projects that LLNL has completed or in which it is actively engaged. Classified activity studies were not included in this project. However, this was not detrimental to this project, because the range of research projects examined included research that augments both nuclear weapon stockpile maintenance and socially beneficial programs. The programs examined demonstrate the LLNL non-nuclear benefits, while conducting research for its nuclear stockpile mission. Therefore, classified work, which represents a small percentage of the overall programs, did not affect this project.

In fact, the work that may be classified has created expansion of education in all areas that are not classified. This effect is referred to in this project as the LLNL paradox. But this does not exclude the fact that some general project information is simply not available, and subsequently not articulated in this study. The general mission of research activities for all national laboratories and controlling agencies are considered. Lawrence Livermore National laboratory (LLNL) is the focus of this research work. The primary activity is to analyze the laboratory mission statement. This is accomplished by examining the engineering sciences research operations. To understand the underlying issues for development of a mission and knowledge and innovation production, the appropriate concepts are studied in depth.

CHAPTER 2

LITERATURE REVIEW

Analysis of the mission of the Lawrence Livermore National Laboratory (LLNL) and the engineering sciences research operations was accomplished as a coupled project, because the overall mission affects the research operations. In addition, analysis of the research operations capability and planning defines an operational mission that is offset from the overall stated mission. A discussion of this point is provided in the analysis of Chapter 4, and the conclusions of Chapter 5. Many interacting forces affect the operations

and mission of LLNL, as indicated by the topics shown in the literature review which follows. Due to the complexity of the forces, the literature review is broken into categories of topics. These are seen collectively as the forces which affect the actual operations and mission. The literature cited includes records from evidentiary government hearings, which show that the mission and purpose for operating advanced research laboratories may be driven by controversy found on a global scale.

The method of this research and study is an analysis of existing evaluation and evidence found in congressional records. Standard regulations from the General Accounting Office (GAO) for evaluation methods are cited, along with pertinent laws, regulations, and policies.

SECTION 1.Science Knowledge and its Management

drawn from advanced research, requires a more detailed analysis of theories of knowledge and innovation. A survey of literature for this perspective is thus included. The literature review sections presented below are included to develop a basis for the scope of the interacting forces in complex research. The topic areas are descriptive, but the integration of the group provides a representation of the interacting forces. The analysis of the subsequent research, which is drawn from multiple perspectives, provides a basis for development of new models for analysis of large-scale and interactive research. The results of the research, toward an appropriate operations model, are defined in Chapter 4.

The two primary areas of concern for Jain and Triandis (1990) are subtle emotional issues. On the one hand, a program that is envisioned to not yield agreeable results can reduce the potential for results. A compounded difficulty can arise from this. If the expected yield is not necessarily high, then a possible plan for the program can be effected. In other words, a non-cooperative start is likely to yield an insufficient program plan—which then insures a higher level of potential failure.

26

Mixed effects can be estimated from the existence of multiple avenues of research. For example: LLNL experiences mixed support from Congress. In other words, the congressional support for each program changes, affecting LLNL planning. Large-scale program changes affects the research planning. Programs which have mixed support from Congress could produce uncertainty of output value. In addition, changing regulations for cooperative research programs could affect the balance of responsibility and expectations of the research personnel.

Given that technology transfer requirements can change, the quality and product completion timing could be affected. The overall effect of a noncooperation atmosphere appears to be a compounded. Although Jain and Triandis (1990) offer significant evidence for the effects, the program reports and productivity of LLNL (that are listed in the Technology Assessment of Chapter 4) appear to have overall high productivity.

From the Jain and Triandis (1990) conclusions, a small increase in congressional cooperation may yield a compounding growth effect on an individual research activity. In addition, a change in a required cooperative research program may produce a lag in a cooperative atmosphere. Therefore, a rapid change requirement may affect efficient development in a program.

Analysis of global research activities provides a broad view of impacts on national

research planning. This broad view can be drawn from consideration of research in

Europe. Henry (1991) presents a survey of technology development in Europe. This

book is part of a series of classes offered by the European Community. These courses are

software and telecommunications offerings entitled "EUROCOURCES". Authors for

each country represented are referenced. Henry (1991) is the cited author. A collection of

technology evaluations for the European community is presented. Henry begins by

stating that "both forecasting and innovation are highly stochastic

Supported by careful evaluations and operational planning. Henry gives an extremely

broad basis. The entire European community is given as an appropriate planning basis for

the best innovation production. In presenting the hypothesis that innovation prediction is

stochastic, one finds that use of standard business analysis tools may not be appropriate.

Henry (1991) claims that prediction is also the sum of two low probabilities. This

suggests compounded variables and, therefore, a set of simultaneous differential

computations may apply as well.

Therefore, the prediction of innovation would be a somewhat complicated analysis adventure. But the balance of the book by Henry (1991) suggests methods that can remove some of the variables. This is offered by means of adapting marketing analysis methods. But this adaptation can represent stochastic processes as well, given that Henry notes a stochastic process for predicting innovation. The significance of the book rests on the presentation of planned detailing of variables, in order to discover and eliminate some variables. This suggests that the primary methods needed require diligence in research and analysis, when applied to research and analysis.

Henry's (1991) overall assumption reveals a form of conflict. It is well known that United States research activities regularly produce patentable innovation. Furthermore, this is apparently accomplished while there has been limited or mixed support from Congress. The existence of many patented ideas is evident in the LLNL project evaluation data shown in Chapter 4-Part 2, and shown in Table 1 of this project. But this is accomplishment contrasts with Jain and Triandis (1990), projections, as they suggest that non cooperation will reduce output. At issue is the point that Henry (1991) does not include an effect from noncooperation, even though an across-borders analysis is part of this marketing plan.

A significant difference exists in the planning needs of the European community. They must cope with rigid national boundaries and broad cultural differences. But the U.S. must cope with state boundaries, and congressional priority changes. Thus, in considering the issues of cooperation defined by Jain and Triandis (1991), a prediction of noncooperation resulting from social boundaries of any kind may give some validity to the Henry's (1991) prediction that innovation may not be predictable.

Fusfeld and Haklisch (1984) provide an analysis of innovation issues with regard to a varied form of laboratory or research structures. Their book is based on an international conference on university- industry research interactions. The authors report complex difficulties in assessing and determining effective practice in research. They suggest that "The university should be the center where really new concepts are bom and where really new phenomena are discovered" (book preamble). They note that the proper ingredient for success is "competence—not money." In addition, "competency represents an organization quality" (p. 13). Fusfeld and Haklisch hypothesize that "The independent, but related paths of university and industry seem to draw more closely together, as the influence of technical change throughout society increases" (introduction).

Issues about knowledge dominate the book. This is shown by consideration of the topics: "What is knowledge?" and "Who is responsible for the generation of knowledge?" Standards for technical product quality development are considered. The authors propose that universities need to be part of national standards development, and present the German DIN system (German standard product system) as a good example of national standards that have led to increased market success (p. 166).

The authors also present a collection of ideas for improving the effectiveness of university-industry R&D, and define the items for improved cooperation. One key consideration is that science parks be located in proximity to universities. Another is a general acceptance that education needs to be a life-long process, and the industry-university blended cooperation can make this more of a reality (p. 162).

This concept is given as an argument of the hypothesis the university be the center of research. Because, the issue of knowledge growth may not present the same boundaries and timing issues that may otherwise be presented. But a definition of knowledge and its effective use are at issue when considering articles from the contributing authors. For example:

The contributing author Nicolin (1984) suggests that: "To make use of knowledge we need an active process. We need willpower. We need a goal or a direction in which to move. Would it be correct to describe active knowledge as competence? But competence requires even more information. Any process where knowledge is applied reties on information. We used to regard competence as a quality of man, but it is equally a quality of an organization.... They are then combined with the aim to minimize the cost of the end product. Should, for any reason, the relative process change, then another combination becomes optimal. The same holds true; should any of the production factors change the quality at a given price. Hence, knowledge affects the impact of production factors, in the industrial process" (p. 15).

The contributing author Bugliarello (1984) reports that knowledge is the elusive value. Perhaps the core of the problem is that-more than two thousand years after Plato, and in spite of Descartes, Kant and Chomsky-the concept of what knowledge is continues to remain elusive (p. 16).

There are elements of agreement between the authors of this book and Henry (1991). One key comment is that "it is competence, not money, that produces the best results." Henry implies this by determining that prediction of innovation is nearly impossible. It is argued by this researcher that the reason for difficulty in innovation prediction rests on the issue of the definition of knowledge and its production, because innovation represents evidence of knowledge growth. Therefore, research toward knowledge growth (and subsequent innovation) would be hampered if the definition is not well understood. One interprets this as a functional agreement of two large groups of authors that represent both the United States and the European community. One summarizes the findings in all three of these references noted thus far. This is that intensive research, and improved planning of research, are fundamental needs. In addition, a need exists for a more clear understanding of knowledge growth. One might argue that a comparison and consideration of the effects of research agency missions must, at its core, consider efficient knowledge production and dissemination. Yet Bugliarello (1984) begins by stating that the definition of knowledge is elusive. But Fusefeld and Hacklisch (1984) are arguing on behalf of the value of university-centered research: "That universities should be the centers where really new concepts are bom." How could they conclude this is the most efficient source for knowledge development, if there is only an elusive definition for knowledge?

The Bulgiarello (1984) entry on the issues of sources of knowledge is helpful for consideration of this project, particularly, since this author claims that "in spite of Descartes, Kant and Chomsky, the concept of what knowledge is remains to be elusive". I

In addition, the Nicolin (1984) article suggests there is difficulty in decisions on how to use knowledge once it is found. These are seen as fundamental issues for analysis and development of a national laboratory mission.

Key concepts for knowledge production are explored in the Concept Fusion model (Cutburth, 1997) A significant expansion, with regard to the teaching of Kant (1781), is delineated in the Concept Fusion model, and is outlined in Chapter 4 of this work. Because national laboratories are normally valued for their development of knowledge through research it is argued that a mission should center around knowledge production.

Monck (1988) provides another large-scale perspective on research issues. The research management structure is shown to have been impacted by World War II. Monck begins by noting the history of industry-university interactions. He compares forms of consortia, specifically Govemment-University-Industry, University-Centered Research Cooperatives, and Direct Company-University research interactions.

For the Government -University-Industry approach, Monck (1988) describes the
needs of the Second World War. Military developments had not kept pace with the
technology found in the Axis powers. Government funding grew. This was followed by
the Cold War and technical threats from the Soviet Union particularly when the atomic
bomb secrets were given to the communists (p. 7). He reports that continued government
influence caused PhD's to become more aloof and reject industrial employment (p. 7).
Large-scale space programs continued this trend. Large-scale funding was given for
government research, but only small amounts found their way to industry. Monck reports
that "This created a relatively ineffective way to transfer technology. The matter was
further exacerbated by

Washington's chaotic and restrictive patent and intellectual property rights policy
regarding government supported research" (p. 7). Monck (1988) describes more problems
for U.S. competitiveness caused by the prior plan. This is followed by an outline of the
university-centered research cooperatives and consortia. He notes the Massachusetts
Institute of Technology plan started in 1980, and reports that "the number and variety of
shared research structures have proliferated. This trend continues unabated. There are
other difficulties, in spite of this improvement" (p. 35).

Monck (1988) warns that there are difficulties for any consortia in the area of patents and licensing agreement. He warns further that "many industries are skeptical because large financial loses can be experienced" (p. 47).

Comparison of university research from industrial grants is given. However, indirect expenses are noted as a constant irritation. These are funds and expenses that occur during the research. "Government and Industry are continuously finding difficulty with the university fund handling" (p. 59).

Monck (1988) offers that in spite of the problems, project planning is a key to higher levels of success. Programmatic strategies are noted as primary to the general objectives and strategies. He reports that a technology roadmap for getting started is needed. The planning suggests prior analysis of the mission, analysis of the key technology areas, the core competencies of the research actors, and available facility for the technology mission" (p. 63). Monck (1988) does not claim that evaluation and planning will produce innovation. However, he argues that a detailed analysis methodology may produce some improved cost control. Furthermore, the explicit plan to evaluate core competencies is in agreement with the needs noted earlier by Fusfeld and Haklisch (1984).

Monck (1988, appendix) provides a few sample research contract outlines. These sample contracts emulate real legal contracts which could be copied. This is used to help define and perhaps reduce some of the common problems from oversights that occur. However, social communication issues could be predominant, and present subtleties that are not found for a contract. Because the organizational competency and issues of cooperation noted above may not be apparent in a contract arrangement.

Furthermore, if a definition of knowledge is not clear and subsequent low probability for innovation exists, the drawing of a contract instrument could be considered superfluous. A contract for monetary and programmatic issues would be needed. But the consignors would perhaps find greater value if a final product was to be envisioned. This suggests that higher order of analysis systems are needed which can resolve these issues.

Monck (1988) suggests that almost any plan will provide benefit if there is adequate and appropriate accounting and analysis. However, the approach does not address the issue of education by means of research. Fusfeld and Haklisch (1984) suggested that the universities need to take some form of lead position to insure adequate education. But Fusfeld and Haklisch also reported cost overrun and control issues. It is argued that this will continue whether there is a contract or not, when the underlying issue of a lack of understanding of the definition of knowledge and innovation production exists.

Considering the Fusfeld and Haklisch issues, the Monck plan for general evaluations would not guarantee a correct interpretation from his competency analysis. Summerton (1994) provides another perspective of research planning in the European community. Her book uses systems approach concepts for considering large-scale change, and notes case evaluations for major changes in the European community. She provides an historical analysis of technology develop the idea that large-scale change can happen, particularly when considering the economics and wars from 1890 to the present.

Summerton (1994) defines technology as a system. "The point of departure for a systems perspective is that many technologies cannot be viewed as isolated artifacts. Instead they are parts of larger 'wholes'. For example; The telephone is seen as a household artifact. But, in reality it represents worldwide telecommunication system" (p. 3). System and systemicy are considered at length, with conditions of the difficulty to separate interconnectivity for any technology. A definition of systemicy is provided: To gain admission to a web of complex sociotechnical systems. To buy a car is, in a real sense, to buy into complex road, energy supply, parts distribution, maintenance, registration, insurance, police and legal systems....Implicit to the systems approach is thus the perspective that in seamless webs of technology, the 'technical' and 'social' dimensions of technology are intertwined. The technical is inherently social (p. 3).

Summerton (1994) asserts that technology has cultural embededness, and the potential for change is affected by culture. The systems suggested consider the across boundary conditions. IBM and Siemens are given as examples of corporations with international operations that were made possible by the development of advanced communications systems.

The book contributors, Braun and Joerges (1994, p. 25), define methods for recombining large

technical systems, also giving examples from the European community. They describe the Trans- border Organ Transplant Systems (TOTS). This European change analysis method is given as an all-encompassing evaluation of the ideas, "The crossing of borders—and attendant blurring of categories-between body and machine, gift and commodity, individual and collective ownership, moral duty an abomination, death and life" (p. 25). Therefore the Summerton approach appears to have both systemic and holistic social components. This approach also suggests the existence and understanding of the complex cultural differences that the European community faces. The TOTS plan is explained to be a planned spreading out across borders.

Braun and Joergers (1994) explain that, historically, large corporate structures begin in rather erratic fashion. They finally can operate across border organizations. But TOTS is designed to be a planned top-down spread.

Research funding was applied to effect the experimentation, using medical technology center development—across borders. This allowed more rapid adaptation of newly discovered medical techniques. But the issue began with not having appropriate centers for the adaptation of surgery requiring high technology equipment.

A second-order evaluation need is presented. This is called the "second-order large technical system "which includes consideration "interdependent technical heterogeneity" (p. 42). The graphic descriptions for the overall evaluation concepts are shown as a network cabling system, with all of the internal branches tied to some other branch. This graphic picture is intended as a metaphoric representation of interdependent research and product connectivity (p. 46).

A clear point made by Summerton (1994) and contributors is that consideration of the overlap and connectivity of research operations components is a fundamental starting point for a strategic plan. Braun and Joerges (1994) go further by noting that a complete technology analysis should include all aspects—even death and life. This is given as consideration for the marketing analysis that includes consideration of cultures and political boundaries. Summerton's book seems to align with the call of all of the prior cited books, in which are found a continuous need for completeness, connected, and broad analysis considerations. But, all aspects are considered, when all of death and life, individual and collective ownership, and cultural diversity are noted in the book.

Questions are raised by Toffler (1990), noted next, with regard to the consistency of

most plans. Simply to write in a book that all issues matter, even death and life, for an

analysis, does not guarantee that the boadest perspecive will be considered, or that it will

accomplish the intended results. Toffler (1990) gives analogies to describe current,

worldwide, and dynamic change. He provides a topic heading, God-in-a-white-coat, to

describe the past medical profession. He reports that prescriptions used to be written in

Latin, and medical material reading was restricted. He notes that the information age and

new technology has opened international boundaries (p. 7).The zigzag of power is shown

for the growth of the global economy. "The financial marketplace itself becomes so vast t

that it dwarfs any single institution, company or individual" (p. 51), Post-Wall-Street

is the description for a "not-so-viable entity" (p. 51). The Spectrum of Mind-Work is

explained as the new perspective on work functions. The change in work and research

activity is noted by stark comments by Toffler (1990):

The super-symbolic economy makes obsolete not only our concepts of

unemployment, but our concepts of work as well. To understand it, and the larger

struggles that it triggers, we will even need a fresh vocabulary. Thus, the division of the

economy into such sectors as agricultural, industrial, and manufacturing and services

today obscures, rather than clarifies. Today's high-speed changes blur the once-neat

distinctions. Instead of clinging to the old classifications, we need to look behind the labels and ask what people in these companies actually have to do to create added value.

Once we pose this question, we find that more and more of the work in all three sectors consists of symbolic processing or mind-work (p. 72).Info-Tactics*. Today we live in an age of instant media. Yet the more data, information, and knowledge are used in governing as we penetrate deeper into the information society, the more difficult it may become for anyone—political leaders included—to know what is really going on. Much has been written about how TV and the press distort our image of reality through conscious bias, censorship, and even in inadvertent ways. Intelligent citizens question the political objectivity of both print and electronic media. Yet there is a deeper level of distortion that has been little studied, analyzed, or understood. In the coming political crises that face the high-tech democracies, all sides—politicians and bureaucrats, as well as the military, the corporate lobbies, and the swelling tide of citizen's group will use Info-Tactics . Like military force, economical clout is based on knowledge. High technology is congealed knowledge. As the super-symbolic economy spreads, the value of leading-edge technology soars (p. 259).

The primary research planning recommendations given in all the prior references seem to provide a validation of Toffler's (1990) projections. Given the market and planning strategies offered in the other texts, it is understandable that the rate of change and information needs are accelerating rapidly.

This is particularly true when global or across-border planning is developing. The breadth depth and interconnected analysis needs are accelerating One other point is given. One must examine the actual work function need, not the name of the company, or the mission. Toffler's (1990) spectrum of mind-work" suggests the need to have close connectivity and a high degree of communication for university and industry relations.

Fusfeld and Haklisch (1984) suggest that this should be the preferred plan for the future. Toffler (1990) notes change potential that goes beyond those suggested by the above authors. Furthermore, questions of information quality and accuracy were shown to drive a form of defense by entrepreneurial individuals. This suggests that across-border and large-scale planning may still find significant opposition by cultural entities. Furthermore, a question of cooperation arises when there exists social need to be independent and defensive. Jain and Triandis (1990) write that cooperation is essential for achieving optimal results.

Therefore, the viability of optimal across-border efficiency can be at issue where defensive information planning is used. Similarly, the Braun and Joerges (1994) inclusion of the consideration of death and life in analysis suggests a great deal of research will be required to overcome the communication issues defined by Toffler (1990).

Carboni (1990) provides historical perspectives on innovation. He begins with information about trends in research, development, and technology, and appears to agree with the ideas of other authors on the unpredictability in research. He writes: One of the major trends in the 1980's to 1990's will be the transformation of an industrial society into an informational society. The central economic fact about the processes of invention and research, is that they are devoted to the production of information The generation of information requires research; therefore, research is going to be one of the most important jobs in the society of the 1990's. Researchers will be the central players in the future, especially in the advanced industrialized economies. In addition to the R&D organizations focus on information, the work itself involves considerable uncertainty since the output can never be predicted perfectly from the various inputs used, (introduction)

Carboni (1990) suggests that the technology and the information age have transformed society, and predicts that technology development will be central to management needs. This suggests that social life and language will be transformed. Therefore, science-based language will become more predominant. Examples are found in the new computer language concept.

Computer literate means to understand the use and the language of computers. However, it is suggested by this researcher that simply understanding that there are changes will not be sufficient to meet the specific needs that are developing. The concept of language expansion or evolution should be considered when examining the balance of the text and journal articles.

In a sequence of structured chapters, Carboni (1990) offers a rationale and methodology for planning and managing of research and development. Only part of the list is entered here. Carboni writes about R&D organization and categories. These include the elements needs, as in the people, their specialization, communication networks and fit of the person to the job. The topics are essentially about the organizational structure, the people within it and their environment.

Carboni indicates that most management and planning needs can be incorporated in a general corporate strategy. A significant difference is that the R&D components may not always be a central driving factor for analysis of potential organizational change. It is suggested that the organization that does not use any R&D should be cognizant of technology changes that can develop. The chapter topics suggest detailed planning to help reduce possible difficulties. For example; "planning for managing conflict" suggests reduction of problems. This would align with the comments by Jain and Triandis (1990) about cooperative efforts. Therefore, rather than studying the phenomenon of non-cooperation,

Carboni (1990) suggests development of a preplan, assuming there may be conflict to handle.

One of the key issues with regard to R&D centers around the ability to turn the work into a product that will create a profit, or provide some benefit to society. The development of the information age, along with new computer systems, has caused significant changes. Thus, to ignore the potential for future change would not be advisable. Certainly, computer systems changes affect banks. Planning for financial strength should include evaluation of the technical base on which an industrial loan is based. Of course, history shows, with all of the bank and S & L failures, there does not seem to be a great deal of analysis going on. In addition, since Toffier (1990, p. 51) questions stock market instability, this suggests economic difficulty in many categories that may be caused by failure to understand the broad problems that exist. (MOT) is offered in the handbook edited by Gaynor (1996). This handbook provides material from authors in 38 chapters. A critique of the handbook is outside the scope of this project. However, key items are surveyed with respect to a focus on MOT and Research and Development (R&D).

Gaynor (1996) notes the opening statements in a 1986 conference on MOT. He writes: "The articles suggest that the failures of the automotive, office equipment, and electronics industries were not a result of trade, economic, or political policies. But, from the inability of industry to implement programs in technology management" (p. 13). He reports further that one major concern was that "universities could be accused of false advertising in their depiction of MOT courses that there is seldom any teaching of either engineering management or MOT. This situation is not unusual; academic institutions rarely recognize interdisciplinary study of research" (p. 13). Given this broad statement, the handbook appears to present concepts that offer adaptability for education in MOT. However, it seems to present a vividly show that the education of MOT, up until this 1996 handbook, may be seriously lacking in depth or adaptability. This would suggest that needed concepts are not generally available to managers toward MOT. Furthermore, it suggests that a sort of update is needed for many. The MOT handbook, chapter 5, provides a sense of the changes occurring. A contributor (Van Wyk, p. 5.2) writes regarding "The Corporate Board and the Need for Technology Analysis," "It is not the tradition of the corporate board to be involved in MOT....this is no longer the case. Corporate boards are rethinking their roles." Van Wyk surveyed corporations to discover the climate and found that "Almost two out of three board members interviewed, identified 'setting the strategic direction of the company' as one of the jobs" (p. 5.3).

The Chapter 5 Technology analysis, foundation for technological expertise, includes

tips on process list which includes the following items: describing, classifying,

identifying interactions, and tracking change of technology. This analysis is focused on

charting technology breakthrough zones. In addition, the author suggests there are levels

of detail that can be added for an information structure for the topics (p. 5.11.2).

Chapter 4 examines the technological base of the company. The author begins by

noting that "Technology-based organizations are constantly confronted with dynamic

and unpredictable changes in markets, products, and technology." The key elements are

complementary assets, organizational assets, and core technological assets, to develop

products and services these are shown to include adaptation of external assets. This is

suggested to consider procurements where necessary to fill the needs of a plan (p. 4.2.3).

In Chapter 12, the author suggests methods for forecasting and planning

technology, and varied approaches of forecasting. He suggests that there have been three

approaches for forecasting technology: "(1) extrapolation of past and current trends, (2)

structural analysis of underlying factors determining these trends, and (3) exploration of

possible changes in the structural factors" (p. 12.1).Betz (1996) includes a

"morphological analysis of a technology system." He also suggests that planning can be

based on morphological analysis. This is defined as an analysis of the components and their relationship in a product (p. 12.9).

In Chapter 23, A framework for product model and family competition is offered. The author focuses on business entities which are able to produce models and families of products. Automobile and home product families are compared using life cycle performance analysis (p. 23.2). In Chapter 28, the contributing authors Sherman and Souder (p. 28. 1) introduce factors influencing effective integration in technical organizations. The authors begin with notes about the necessity of planning for technical integration in a research organization: "In any complex new-product development project, the effective integration of diverse inputs from cross-functional groups is important in meeting schedule and budget requirements" (p. 28.1). Integration is often critical to the success of the new product or system. Small systems are mentioned, but large programs are noted here. The author continues

"However, with large-scale projects in which there are many interdependent subsystems and coordination is required across functional areas or organizations (i.e. contractors and subcontractors), the issues of integration become more crucial" (p. 28.1). Sherman and Souder (1996) compare organizational structures for integration. Functional matrix structures, balanced matrix structures, and project matrix structures are mentioned. The definition is shown to be a focus on the management structure.

The combined tools of Gaynor (1996) and Honig (1986), suggest that technology

exists which can be used to manage technology. Furthermore, they demonstrate that

more complex and highly complicated programs can be managed with varying degrees of

detail. It is argued that, in spite of the detail of analysis possible unpredictability of the

impact of technology, and the variability of management that corporation or some

research agencies may exhibit, will allow a continuation in large-scale failures.

The articles demonstrate that improvements are occurring in management tools,

except that adaptation of the elements and collection of methodologies are subject to

unpredictability. In Chapter 4 of this project it is shown that LLNL uses a form of matrix

management. However, it is demonstrated that standard business analysis methods do not

offer the complex computation systems that are needed for a complex research agency of

this type. The unpredictability described by the above authors is considered as

demonstration of the stochasticity of human action for which environmental computation

methods are needed. The need for complex differential computations methods is

demonstrated below. Therefore, although the above authors suggest methods for

assembly of complex information, they do not show that these methods represent

complex stochastic differential systems computation needs. In other words, if there is an

unclear definition of knowledge, then this issue suggests that its creation, acquisition, and dissemination is nonlinear.

Henry (1991, Introduction) argues that the prediction of innovation is the product of two low probabilities. Whereas this can be a basis for considering stochastic processes, the authors did not concentrate on this issue for defining a clear solution and guaranteed success in research. This researcher argues that the most significant basis of the difficulty rests on the definition of knowledge, and its relation to innovation production. Therefore, it is argued that this must be a significant starting area for improved understanding of knowledge growth for an ideal research agency. Key concepts of understanding knowledge and innovation center on the theory of measure, and the existence of pre-existing knowledge potential (known as a priori knowledge). These concepts are briefly noted by Fusfeld and Haklisch relate to a theory of knowledge offered by Kant (1781).

In addition, their relation to advanced and modem mathematical game theory are considered by Von Neumann and Morgenstem (1946/1972). This classic text defines methods for analyzing conflicting strategies. One sets conflicting requirements in the form of a game, and uses the cross comparisons to help optimize an approach that is likely to win. This text is cited repeatedly in texts for operations research and management sciences. One can use the methods for political and economical choices. To consider the principle of knowledge production and innovation, one examines the issues of mental measurements for examining the existence of the knowledge.

Von Neumann and Morgenstern (1946/1972) detail levels of complexity in mathematical game theory. But, as the title of their book indicates, Theory of Games and economic behavior, there is relation of the situations that exist in economics to the concepts of the creation of knowledge. Games of strategy and analysis are used for studying the possibilities of choices. But human characteristics present

broad variations in potential results. For example: They delineate a definition of economic utility. This is suggested to follow the psychological parameter of human characteristics with regard to choice and preference. In other words, the examination of a theoretical game strategy for complicated processes should begin in this large perspective. They compare the analysis of variability of choices to the development of the theory of heat, its measure, and computation. On the notion of utility-. Historically, utility was first conceived as quantitatively measurable, i.e. as a number.

Valid objections can be and have been made against this view in its original, naive form It is clear that every measurement—or rather every claim of measurability—must ultimately be based on some immediate sensation, which possibly cannot and certainly need not be analyzed farther. In the case of utility the immediate sensation of preference—of one object or aggregate of objects as against another—provides this basis.

All this is strongly reminiscent of the conditions existent at the beginning of the theory of heat: that too was based on the intuitively clear concept of one body feeling warmer than another, yet there was no immediate way to express significantly by how much, or how many times, or in what sense. This comparison with heat also shows how little one can forecast a priori what the ultimate shape of such a theory will be. It turned out to be additive and also in an unexpected way connected with mechanical energy which was numerical anyhow. The latter is also numerical, (p. 15)

This theoretical examination of notions of economic utility, the creation of a theory of heat, and their relation to mathematical games of strategy, presents a foundation of the issues under study in complex research activity. One argues that their analysis methods provide a unique assessment for the quest for a theory of measure for all things. This provides a basis for comparative analysis for complex research. This compares to the issue of the quest for a definition of knowledge and its production, as found in the book by Fusfeld and Haklisch (1984). They concluded that a definition of knowledge is yet to be completely resolved

Von Neumann and Morganstem (1946/1972) delineate mathematical theoretical game analysis for the purposes of understanding the effects for planning in complex situations. They suggest that, where preferences-choices are concerned, complex computation systems are involved.

They explore the stochasticity (which is probability in a complex differential domain) for complex multiple choice issues. The issue of Von Neumann and Maorganstem definition of the beginning point about measure and its theory should be considered with regard to LLNL's research on instrumentation for measurement of complex physical processes. LLNL shows proficiency in that area, as noted with regard to their report UCRL-TB-119860. But the primary issue with reference to this text is that complex research requires complex analysis of subtle notions of measure, with regard to the concept of knowledge, and innovation production.

The adaptability of the concepts presented by Von Neumann and Morganstem are found in current texts. For example: Houseman (1977) provides textual instruction for quantitative analysis. The methods listed include elements for Linear Programming, Game Theory, Simulation modeling, Queuing theory, and Markov Processes. A growth and institutionalization of mathematical game theory, simulation modeling, queuing theory are demonstrated in this text. A consideration of utility is provided. He writes:

Is it possible to measure the utility of money? In an attempt to answer this question, we shall consider three different types of measurement scales: 1. Nominal or classification scale, 2. Ordinal or ranking scale, and 3. Cardinal or interval scale. The Von

Neumann and Morganstem measure of utility is a special type of cardinal measure (some would say it is a special type of ordinal measure).

The use of this utility measure allows us to predict which of several lotteries a person will prefer. One justification for the use of utility in investment situations evolves from the fact that other measures cannot adequately cope with uncertainty (p. 198).

Since human action models include differentiable variables, stochastic process theory is to be found in more of the advanced texts, as in this one. Game theory models can be defined as linear or non-linear. The Houseman (1977) choices for modeling center on non-differentiable games. But the issues of human action for utility are examined by use of the concepts from Von Neumann and Morganstem (1946/1972). A business analyst should be cognizant of the issue that variabilities exist where complex human choice issues are concerned. It is argued that evaluation methods should include elements of non-linear models.. However, although game analysis is used in the text, it is shown that the application for complex research is non-linear, and that stochastic differential systems are present. This point is demonstrated in Chapter 4 of this project.

The complexity of analysis in advanced research is demonstrated further when examining the interaction of technology and its supportive mathematical concepts. It is argued that the actual mathematical analysis requirements are more complex than what may be regularly considered.

Kloeden and Platen(1992) demonstrate that complex analysis in mathematics is

somewhat generic, and adaptable to multiple categories of research analysis

simultaneously. Thus, when examining knowledge growth and subsequent innovation

production, one may use the rationale and methods found in theoretical mathematics.

This issue is examined further in Chapter 4 of this projecE delineate theoretical

mathematical stochastic differential systems. They provide definitions of basic principles

for a definition of probability, and provide software for analysis and solution of stochastic

differential equations. This in turn relates to the concepts presented by Von

Neumann and Morganstem. Kloeden and Platen (1992) report that these theories

began in theories of measure. They write: The axiomatic development of probability

theory (where uncertainty exists) was initiated by Kolmogorov in the early 1930's. It is

based on measure theory, but it has developed characteristics and methods of its own.
The

fundamental concept in this approach to probability theory is the probability

space (p. 51). Kloeden and Platen (1992) provide notes on applications of stochastic

differential equations. For example: formulas are delineated on human population

dynamics, protein kinetics, genetics, experimental psychology, neuronal activity (neural science), turbulent diffusion and radio-astronomy, biological waste treatment, hydrology

and indoor air quality, seismology and structural mechanics (and other areas).

There are three essential points of this brief quote and list of application areas. The

first is that the principles of mathematical analysis begin with analysis of measure, using

theories of measure. It is argued that this should be compared to the issues delineated by

Von Neumann and Morganstem (1947/1972) for theory of measure. This should be

compared to the issues from Fusfeld and Haklisch transformed. But stochastic

differential computation systems do apply in analysis of complex operations.

However, the collection of computation methods shown suggests that mathematical

computation systems are somewhat generic, and offer broad analysis application

potential. The essential issue argued for this project is that most models are not fully

developed for analysis of a complex research organization. Each author provides helpful

ideas which can be integrated into a larger system of analysis.

Saaty and Alexander (1989) define and develop mathematical analysis tools for

decisions. These tools are constructed around ideas of how humans appear to rationalize

their environment and their decision issues. A primary concern of the book is to examine

the analysis of a hierarchy which is present in this process. They explain: In solving

problems by explicit logical analysis, three principles can be distinguished: the principle

of constructing hierarchies, the principle of establishing priorities, and the principle of logical consistency. Structuring Hierarchies : Humans have the ability to perceive things and ideas, to identify them, and to communicate what they observe.

For detailed knowledge our minds structure complex reality into its constituent parts, and these in turn into their parts, and so on hierarchically. The number of parts usually ranges between five and nine. Setting Priorities: Humans also have the ability to perceive relationships among the things they observe, to compare pairs of similar things against certain criteria, and to discriminate between both members of a pair by judging the intensity of their preference for one over the other. Then they synthesize their judgments—through imagination or, with the analytical higherarchy process, through a new logical process-gaining a better understanding of the whole system (p. 17).

Saaty and Alexander (1989) draw these fundamental thinking process steps into mathematical relationships for computation in order to assist in decision making. The teaching is essentially to help one focus on decisions where there seems to be a conflict. Matrix computation methods are used, for pair-wise comparison of data which is picked, with regard to the conflicting decision issues. The matrix of decision issues are expanded where there are more complicated information elements which add to the complex decision issues. However, this must be understood to be information which is drawn from seemingly rational thinking. Most humans believe they are using rational thinking at all times

Therefore, matrix computation, developed as simultaneous differential and stochastic computation systems, may be used for examination of variables that include some human emotion considerations. See Von Neumann and Morganstem (1946/1972), Basar and Haurie (1994), Dresher(1981), Schrage (1997), Kosco (1994), and Churchland and Sejnowski (1992).

The Saaty and Alexander (1989) matrix computation methods are written into software that is offered by Expert Choice (Saaty, 1983/1995). This software is used for experimental analysis of decision issues which are based on a construction of priorities. The methods are shown to be useful for helping one define difficult decision ideas in terms of mathematical computation means. However, the software does not include the computational parameters that are with regard to complex science research program analysis.

The software design and computation system assumes that all of the evidence for rational thinking is available to be loaded into the computer. The software is used for an experimental analysis of a set of goals regarding LLNL's operations planning (see Chapter 4 of this project).

The thinking process examination by Saaty and Alexander (1989) and subsequent computation tools help one examine a discrete decision issue, where otherwise mathematical relationships were previously not understood. However, complex science research for knowledge production presents variable computation needs which a discrete analysis is not designed to accomplish.

Vast changes are suggested by this researcher that could assemble ideas from Von Neumann and Morganstem (1974/1972), Basar and Haurei (1994) and Dresher (1981) for experimental analysis of knowledge production and innovation computation. It is argued that the Saaty and Alexander (1989) fundamental concepts of seeking to find a hierarchical principle and variable adaptability actuall demonstrate potential for advanced analysis which the mathematical analysis does not always support. In other words, the computation focuses on a discrete decision issue, but the analysis of human mental processes are regularly thought to be representable by stochastic differential systems.

Therefore, the unique value of the work of Saaty and Alexander (1989) is shown to be in the principles of thinking—to seek an understanding of the hierarchy of thinking. This book and other books that examine concepts of hierarchical thinking, are analyzed further in the Concept Fusion model (Cutburth, 1997) outlined in Chapter 4 of this project.One asks if there can be a relationship in the analysis of science research operations and principles of analysis within nuclear theory.

The answer is argued to be within the principles of mathematics. They are generic to a particular focus. Therefore, as demonstrated in the Concept Fusion model, computer software development of complex mathematical algorithms for adaptation in nuclear physics and associated engineering sciences can relate to the analysis of human characteristics of an advanced research organization, because both represent complex variable systems.

Birdsall and Langdon (1985) write about computer analysis of plasma physics. At the time of pubheation of their book, both physicists were researching for the University of California. Birdsall was at Berkeley, and Langdon was located at LLNL. The text shows considerable advancement in models for analysis, and a growth in plasma physics knowledge. Examples of the computer software are shown for analysis of plasma physics, along with the associated mathematics. Simplified Differential Operator notation is used in presentation of the appropriate computer algorithm systems. Therefore, new models and applications developed for analysis of the processes of physics are shown in the text.

Historical reference is made of the work of James Clerk Maxwell for electromagnet theory. His theories are surveyed in Gray's (1963) Handbook of the American Institute of

Physics: 2nd ed. (AIP).

The key differences are in the adaptation of Differential Operators for computer systems algorithm development, and perhaps streamlined education of the topic. This is due to the primary focus of the text, which appears to be electron plasmas and ionization using basic algorithms. These are shown in the text.

Significant reference and adaptation is presented using Fast Fourier Transforms. T his is taken from the theoretical mathematician, and the Fourier Transform. Although Fourier analysis is found in the AIP handbook noted above, Fast Transform computation has been augmented from computer systems development. Fast Transforms are also used in signal processing and analysis systems. This is shown by Kosko, (1992), in Neural Networks and Fuzzy Systems: A Dynamical Systems Approach To Machine Intelligence. Expanded information about Fast Transforms is found in Elliot, and Rao's (1982). Fast Transforms, Algorithms, Analyses, Applications.

It should be noted that LLNL has been setting a fast pace of development of the analysis of plasma physics and Fast Transform computer software. Other LLNL computer

software development with regard to Fast Transforms is found in reports about the SISAL computation system. (See the LLNL report M- 146). The new National Ignition Facility (NIF), which is an LLNL program, depends creating new objects, which are still at their core, algorithms that are found in computer memory.

It is argued that all three approaches, AFSC, GP, and OOP, each help one develop a greater breadth of knowledge growth potential, due to the combined or aggregated knowledge of the combination. This is examined more fully under the Concept Fusion model presented by this author. Support for the concepts of mental algorithm development is found in the current work for analysis of neural networks. Kosko (1992) provides a technical analysis of neural networks. In addition, an interconnectivity of multiple categories of sciences is demonstrated. The relation of many sciences human brain functions are shown.

Learning theory is shown simultaneously for computer systems, that are adaptable to machines, and to the human mind for acquisition of knowledge. The words related to the brain components are called Neural Networks. He notes that the human brain contains around Eleven Billion neurons, (p. 13). A connected science has grown, because the computer and software designs are made to emulate human brain functions. Kosko (1992) writes:

The Dynamical-Systems Approach to Machine Intelligence: The Brain as a Dynamical System. Several engineering and scientific disciplines study how adaptive systems respond to stimuli. Electrical engineers study the topic as signal processing, nonlinear filtering, coding theory, circuit design, and adaptive control. Computer scientists study it as algorithm and automata theory, computer design, robotics, and artificial intelligence.

Mathematicians study it as function approximations, statistical estimation, combinatorial optimization, and Dynamical systems. Philosophers study it as epistemology, causality, and action. Biologists study it as neuroscience, biophysics, ecology, evolution and population biology. Psychologists study it as reinforcement learning, psychometric, and cognitive science. Cultural anthropologists study it as culture. We shall emphasize electrical engineering as we seek general principles of how adaptive systems process information (p. 12).

Intelligent Behavior as adaptive Model-Free Estimation, Mathematically, all these systems transform inputs to outputs. The transformation defines the input-output function $f : X * Y$. Indeed the transformation defines the system. We can operatively characterize any system-atomic, molecular, biological, ecological, economic or legal, geological, galactic—by how it transforms input quantities into output quantities.

We call system behavior "intelligent" if the system emits appropriate, problem solving responses when faced with problem stimuli. Intelligent systems also Generalize. Hence they estimate continuously . Geometrically, when systems generalize "create", they map stimulus balls to response balls. Consider a known stimulus-response pair (x,y).

Stimulus x defines a point in the stimulus space S, the set of all possible stimuli for the

problem at hand. In practice S often corresponds to the real Euclidean vector space R'

Response y defines a point in the response space R? which may correspond to(pp. 19-20).

Kosko (1992) demonstrates that the research of a large group of sciences have

direct connectivity through, and results in, knowledge growth in the dynamical brain.

Kosko, in later chapters, explains a group of neural network learning criteria.

Mathematical presentations are the basis for demonstration of every process. A

differential vectorial computation basis is shown, with regard to one singular brain

stimulus and respons pair.

Differential operations and stochastic processes are examined in later chapters.

Kosko (1992) suggests that human learning is complex, but also calculable with

sufficient and appropriate models. However, the key issues are shown that the brain

is Dynamic. Basar and Haurie (1994) demonstrate that a Dynamical competitive game

among humans represents a stochastic differential system. Kosko demonstrates that

complete computation methods for intelligent learning demonstrate this level of

computation requirements.

A key point can be seen from the Kosko (1992) book. Kosko demonstrates that most

branches of science are interconnected in analysis when mathematical models of neural

networks are examined. He suggests that changes in the fundamental computations from the specific learning process will transform the computation system, and human knowledge as well. This point is suggested by Cutburth (1981), and is delineated in the Concept Fusion model (Cutburth, 1997), outlined by this author, in this project.

To consider mental algorithm development, a survey of computer algorithm development is needed. Gen and Cheng (1997) provide a general definition of genetic algorithm (GP) production. In addition, they point to the issues of complexity in management analysis when conventional optimization methods are used. They report that the stochasticity of human action creates the need for more sophisticated computer software analysis systems:

Many optimization problems from the industrial engineering world, in particular the manufacturing systems, are very complex in nature and quite hard to solve by conventional optimization techniques. Since the 1960s there has been an increasing interest in imitating living beings to solve such kinds of hard optimization problems. Simulating the natural evolutionary process of human beings results in stochastic optimization techniques called evolutionary algorithms, which can often out perform conventional optimization methods when applied to real world problems The usual form of genetic algorithm was described by Boldberg. Genetic algorithms are stochastic search techniques based on the mechanism of natural selection and natural genetics. Genetic algorithms, differ from conventional search (computer) techniques. They start with an initial set of random solutions called population. Each individual in the population is called a chromosome, representing a solution to the problem at hand.

A chromosome is a string of symbols; it is usually, but not necessarily, a binary bit string (part of a computer algorithm in memory)...The chromosomes evolve in successive iterations, called generations. During each generation, the chromosomes are evaluated, using some measure of fitness. To create the generation, new chromosomes, called offspring, are formed by either (a) fitting two chromosomes from current generation using a crossover operator, or (b) modifying a chromosome using a mutation operator (pp. 1-2). It is argued that the principles of algorithm development are representations or can be interpretations of the mental algorithm productions suggested in the AFSC model by Cutburth (1981). In the developments by Gen and Cheng (1997), the use of the words and concepts of species evolution are to help one understand the process for searches and production of new algorithms, called offspring or mutations.

To fully examine the function of the human mind, current research includes the simultaneous analysis of the human brain. In examining this perspective one finds a connection to analysis of nuclear theory. Churchland and Sejnowski (1992) provide an analysis of human brain functions, but in turn relate this to mathematical functions of the brain. Their book titled The Computational Brain provides glimpse into this connection of the research topics and subsequent potential for algorithm development. The authors present sciences of the brain and its functions.

As in the book title, they demonstrate all of the physical elements and sciences involved in brain activity and show a theory of its computation methods. They demonstrate that brain functions occur at the atomic level and in interacting brain sections. The atomic level of activity is demonstrated in part by their explanation of some of the information output devices called the synapse. They demonstrate that the combined sciences are represented as functions of the brain. These include electronic, photooptic physical processes, chemical processes, and biological elements which all represent elements of systems and hierarchies. Kosko (1992), as noted above, demonstrates the connectivity of a large group of sciences through the connected analysis of neuroscience. Therefore, the scientific mass of the brain, it is argued by this researcher, is able to examine broad and interconnected sciences, due in part to the fact that it represents broad and interconnected science domains. It should be understood that a full description of the brain function processes are outside the scope of this project. However, a brief of brain functions in the area of the synapse, or neuron, is shown here by means of noting two technical terms the authors supply: Ion Channel and Ion Pump. Ion Channel Water-filled pore in a (neuron) cell's membrane that allows ions to flow in and out of the cell (ionic current) according to chemical concentration and electrical gradients. The flow is gated by voltage or chemical binding to associated receptors.

The ionic conductance is a measure of the ease of flow. Ion Pump Integral membrane ATPase enzyme (protein) that uses energy to move ions across neuronal membranes against their concentration gradients. Often, exchanges pairs of ions, e.g., potassium in and sodium out of the cell (p. 466). In the ion pump, an ionized potassium atom and a sodium atom, are exchanged. To be ionized, they have one less or one extra electron orbiting the nucleus. Sodium and potassium atoms are not the same size, but the exchange allows a small change in the electrical energy flow and potential. The change in brain memory and function, therefore, is at the atomic level. This demonstrates that a great deal of detail in human brain function is possible.

Churchland and Sejnowski (1992) describe the computational methods of the brain. As in Kosko (1992), they argue that the brain computes decisions by methods that are describable as differentiable vectors. Connections of many differentiable vecorial computations are shown to be in a matrix format, (pp. 78-79). Thus, from Marsden and Tromba (1988, Vector Calculus), the vectorial computations are interacting and, therefore, describe multiple simultaneous differential computation systems.Returning to the perspective of the mind and current computer algorithm development, one finds

 new topics of research called Object Oriented Modeling. Rumbaugh, Blaha, Premerlani,

Eddy, and Lorensen (1991) provide a general definition and applications of

Object Oriented algorithm development.

This is referred to as OOP above. In addition, they demonstrate that analysis of

strings of information, or object algorithms have connectivity for algorithm search and

development. They write:

Object-oriented modeling and design is a new way of thinking about problems

using models organized around real-world concepts. The fundamental construct is the

object, which combines both data structure and behavior in a single entity. Object-

Oriented models are useful for understanding problems, communicating with application

experts, modeling enterprises, preparing documentation, and designing programs and

databases. Superficially the term 'object-oriented' means that we organize software as a

collection of discrete objects that incorporate both data structure and behavior. This is in

contrast to conventional programming in which data structure and behavior are only

loosely connected.

There is some dispute about exactly what characteristics are required by an

object-oriented approach, but they generally include four aspects: identity, classification,

polymorphism, and inheritance. Identity means that data is quantized into discrete, distinguishable entities called objects (pp. 1-2).

Keeping the right attributes Link attributes. If a property (of an object) depends on the presence of a link, then the property is an attribute of the link and not of a related object. Link attributes are usually obvious on many-to-many associations; they cannot be attached to either class because of their multiplicity.

Link attributes are more subtle on many-to-one associations because they could be attached to the 'many' objects without losing information. Link attributes are also subtle on one-to-one associations (p. 163).Thus, the authors explain the rules for the connectivity and linkage of one object algorithm to another. They suggest subtle relationships. It is argued that, when considering the Cutburth (1981) AFSC model, the subtle mental relationships can be understood on the human emotion issues of the link up of objects and values of thinking, and in development of new thinking about the relationships.

Therefore, a change represents both creation of a new algorithm and knowledge growth, as suggested by Cutburth (1981). The next issue that has developed from the derivation of the new algorithm models is that of the information or algorithm search methods used. Each has unique characteristics, but Gen and Cheng (1997) articulate the mechanism of the search for new algorithms, based on the complexity of human action which demonstrates stochasticity. New models have been developed for other methods for search and discovery of new algorithms.

Returning to the perspective of new technology development from research, one finds that global technology changes are occurring rapidly, and should be responded to in research organizational planning.

Methods of technology development are examined by Hond and Groenewegen (1996). They examine the process from innovation through adaptation and projection into diffusion in society. They report: Technical change is effected when innovations are diffused to its users. The mutual adaptation between technology, and the user's social and institutional context, is an important source of technical change. In the adaptation process, choices are made about the form, the function, and the use of a technology. Therefore, actors in the social and institutional context of technology, have an opportunity to influence technology development. However, opportunities to develop technology along pre-specified paths are largely restricted to the third, and perhaps the second, hierarchical level.In the last case, large R&D programs might provide the basis for a new dominant design, but generally its application takes unpredictable directions.

Thus, the chances for predictable success are small. One interprets the above statement as being somewhat general. Although the level of effect on a technology entry point and its process may effect success, a broadly general statement may not cover all the issues, Because a case by case analysis can yield more clear answers. The measure of the success of any size R&D project should require a value scale for the level or range of success. Furthermore, if an R&D project has not met with success, one can attribute a failure to the management, or a choice in marketing, rather than the science.

Although some product spin-offs are achieved from large programs, a primary direction should not be assumed to be unpredictable. For example: the quest to achieve flight, netted flight. The quest to find replacement for rubber tree rubber, netted derivatives of Neoprene, an advancement beyond the quality of rubber. See Winspear, George G. (1968) The Vanderbilt Rubber Handbook. The quest to develop the home computer yielded exactly that. Therefore, success and application of a large-scale program can have some predicted results when the original objectives are understood. This is suggested by the planning methodology offered by Gaynor (1996), and Honig (1986). It is argued that the significant issue, therefore, rests in the understanding and the detail of the mission, and its planning.

The last form of technology development can be compared to the planned projection and control devised by documentary comments from the national laboratories. In particular, LLNL notes that its objectives are centered on nuclear weapon stewardship, but this would direct the plan and, per the above quoted statement, would restrict and inhibit the net results. The LLNL has a good track record of achievements in the area of a directed focus, as well as other broad areas, as shown in the Technology Assessment. On the other hand, LLNL is involved in development of many technology areas, as is shown in the Technology Assessment of Chapter 4.

Therefore, one finds a possible mixed strategy and results. Hond and Groenewegen (1996) demonstrate the factors that can affect final product development planning. They suggest that advantages can be attained by using the broad perspective of the social effects from technology. This question is explored with regard to the LLNL mission statement, in Chapter 4 and in the conclusions of Chapter 5.

Hughes (1996) describes an R&D management support software system used by the UK Department of Trade and Industry (DTI). The software is used as a planning device to help in the early decision processes for R&D. The journal art appears as a form of "WINDOWS" based software system, using pull-down charts. The software provides a series of screens that interlock and define a planning analysis track for R&D. The windows provide categorical areas for assisted planning—it gives key areas that must be considered in the development of the plan.

FIRST WINDOW-key words are Strategic Analysis, Corporate Mission, Strategic Audit, Competitive Audit.

SECOND WINDOW-key words are Manufacturing Strategy, Assemble Solution

Stream, Challenge Solution Stream, Manufacturing Strategy.

THIRD WINDOW-Key words are Action Planning, Resource Requirements,

Financial Contribution, Prioritize Actions. This is a management support expert system.

Gaynor (1996) suggests adaptation of expert systems, and provides topics for

the information structure. However, although the software suggests a user friendly input

methodology, mathematical models for support calculation systems are not delineated.

One category of the analysis portfolio includes specific manufacturing methodology data.

This suggests close detailing for the initial planning. But theoretical initial

computation systems can be a significant factor of the initial decision processes. This can

provide data for an analysis of the efficiency and precision for a project plan. A

significant issue is with regard to the non existent mathematical analysis of complex

research. However, Hond and Groenewegen (1996) suggest that the program allows

regeneration or repetition and adaptation of a pnor process. This is the unique but

standard attribute of software systems. Taylor and Oats (1996) give practical examples of

knowledge management. They begin with an important reference to consideration in

planning for managing knowledge. The authors report that they have researched this topic

and its application on a group of cases for over five years.

Knowledge Management, regards knowledge as a corporate resource which, if properly managed, can improve a whole range of organizational performance characteristics by enabling an enterprise to be more 'intelligent acting.' Although many organizations understand this concept, they often adopt fragmented approaches to the problem which at best are manifest as Total Quality Management.

Learning Organizations, or Outsourcing and Core Competence development. Other organizations adopt a more technology-based focus e.g. developing knowledge-based systems or intelligent databases. However, Knowledge Management should be a much more strategic and holistic approach, which is now emerging because of several reasons.

1. There has been a growing understanding of the nature and significance of organizational knowledge and its management.

2. The technological dimensions of Knowledge Based Systems are maturing, including the methodologies for system development.

3. Organization-wide computer-based infrastructures are now becoming commonplace in many organizations. (Introduction)

It is argued that significant knowledge growth is dependent on the recognition of concepts presented in this article. But there may be issues about the definition of knowledge. An increase in productivity is suggested in this article, by recognizing and managing knowledge. The authors summarily describe some current techniques for management as being less strategic.Taylor and Oats (1996) present a case study for examination of knowledge management ideas, by presenting an example in

semiconductor technology.

The example suggests, there was little recognition of existent knowledge for the

improvement of the product, thus producing a less than ideal design. It is suggested that

this model concept by Taylor and Oats (1996) could be considered in a frame of

reference that is greater in magnitude than the example, because they provide an

overview of a process affecting the development. The needs for knowledge adaptation

and growth for a national laboratory are crucial for future national security and

advancement.

However, it is argued that Taylor and Oats (1996) may not actually be

representing the knowledge growth that is envisioned by Fusfeld and Haklisch (1984),

since the definition of knowledge is in question. Taylor and Oats (1996)

may be discussing the management of the personnel who poses knowledge. Though the

concepts can allow growth, an understanding of a theory of knowledge can be

differentiated from understanding management of people.A theory of knowledge is

delineated in the Concept Fusion model (Cutburth, 1997).

Harashima (1996) provides an analysis of a new category of research. Prior to the consolidation of a new Journal, (Mechatronics), products were reported in separate journals as essentially mechanical or electrical. In this new 1996 journal, the two topics were deliberate simultaneous in mechanical and electrical or electronic journals.

The journal combines efforts and technology of both the Institute of Electronic Engineers (IEEE) and the American Society of Mechanical Engineers (ASME).This first issue and article describes the primary criteria of the joint technology needs. Specifically, the interface of mechanical devices and their controls is the focus. Prior to this journal, the technology areas were loosely connected. However, many systems failures have occurred in this interface area.

For example: an automobile can have an electrical failure. The actual failure may be caused by mechanical vibration, which causes an electrical device to become dislodged, or a connector to become loose. Therefore the simultaneous analysis of both the mechanical and electrical attributes of a product will provide a more holistic design. For example: the home computer is an electronic device which is also mechanical. The computer "Hard Drive" is a mechanical system, though it contains electronic devices for memory storage and searching. This journal's first issue provides some primary areas of focus for the engineering science disciplines, which are connected in fact. They include Mechanical Engineering, Electrical Engineering, Electronic Engineering, Control

Engineering, Signal Processing technology, and Materials Engineering.

In other words, the overlap and interface needs are brought together in this journal for many engineering sciences categories. The LLNL is involved in research activities of this nature, and this journal composite seems to emulate this form of technology

This journal can be compared to the work by Cutburth (1993) Systems Analysis: How to Assure the Quality of a product. This book suggests that a key factor for the improvement of products is consideration of the overlap and connectivity of integrated technology. This work reports that many failures occur in product design and planning because of failures in understanding technology connectivity needs. An advanced program analysis is recommended for analysis by development of an expert system which is called MEKA analysis. The expert systems software model offered by Hughes (1996) is very similar to that shown by Cutburth (1993). Cutburth suggests the system will help reduce the potential for catastrophic failures. In addition, the integration of multiple categories of engineering sciences is suggested toward that purpose.

The analysis of the LLNL research quality can be considered using the materials noted in the Mechatronics journal. The journal gives data about sensory systems limitations. The LLNL shows core proficiency in sensory systems product development, as found in the Technology Assessment. The articles that follow are a selection with regard to R&D, comparative technology, and technology management. They are shown to provide a reference to new developments that will affect society. For example: The article by Jones (1994) shown next, suggests new developments in protein research. The author envisions that this will generate new protein engineering and subsequent products.

In other articles noted below, definitions of words Smart materials versus Intelligent materials include R&D of correlated developments with regard to biological products, and materials. This suggests that the language of technology is transforming. As

suggested by Cutburth (1981), research creates knowledge growth, and this also affects language. This is explored in detail in the Concept Fusion model.

The idea of language changes which are apparent in new research is found in the example provided by Jones (1994). New words from research are found in the topic of Protein Engineering. Jones (1994) writes about protein structures: One of the major goals of molecular biology is to understand how protein chains fold into a unique three-dimensional structure. Given this knowledge, perhaps the most exciting prospect will be the possibility of designing new proteins to perform designated tasks. The eventual pinnacle of protein engineering will be the fully automated design of a protein with novel structure and function. Achievement of this aim lies far in the future, although some early progress has been made recently. A model development sequence is shown...First build models, then define energy terms, then evaluate models, and matching sequences

(p. 460). This research is not isolated in the UK. Current research in this field is being

conducted through most governments, research universities, and medical supply

corporations. Stock options for the products are offered in every stock exchange,

demonstrating that a knowledge growth item is also a product.

One intuitively understands that the impacts of improved protein manufacture could

cause another round of major changes in the world. Large scale corporate strategy

planners should be cognizant of these developments, and the associated changes in

language. The magnitude of these programs can be studied by data found through the

National Science Foundation, related journals, and corporate newsletters. Volkel (1996)

presents new technology which also helps to show the development of new languages, in

the category of micro optics. He describes microlens array imaging system for

photolithography, and writes:

A Swiss research group writes about the product development research for

photolithography systems improvements. The equipment is used for process systems for

development of computer or photo device microchip manufacture: They investigated

micro-lens array imaging system was developed in connection with a new contactless

photolithography technique called micro-lens lithograph. This new technique is aimed at

flat panel displays, micro-mechanics, or micro-chip module manufacturing that requires

only moderate resolution, on the order of 3 to 5 microns. (One strand of hair is about thirty times as large as one micron). They claim: Conventional imaging systems, such as lithographic or photographic objectives, consist of different optical elements in a linear arrangement. The image is transported within a single optical channel (p. 3323).

This research is toward improved chip quality, density, and overall manufacturing throughput. Therefore, this new level of technology for the field may translate into a new round of advanced home computer systems development. A business strategy should include some analysis of the ongoing technology changes for this category. A few key words are present here. Photolithography, optical array, microns, mask, multi-chip. These words are part of the new technology language. Sychugov, (1996) describes micro optical systems and control systems. He writes:

A Russian research group in Moscow and a researcher in a German university, combine efforts for this project. They write: The planarization of the access to and from an optical waveguide by means of a grating coupler is a necessary condition for the viability of integrated optics technology in non communication applications...They add: This should lead to low-cost and more reliable manufacturing, and mounting of sensors and micro-systems such as CD pick-up heads, displacement interferometers, and biochemical sensors, to mention just a few possible applications (p. 3092).

The above article mentions new products that effect many other products. One point is that the development of one area of research allows product development in other areas. For example, the technology reported is usable in both displacement interferometers, and biochemical sensors, "just to mention a few possible applications". Another issue is that the researchers are in Russia and Germany.

The last article was about improvements in micro-chip manufacturing devices by researchers in Switzerland. An intuitive approach may suggest that these new developments do impact US research needs, and will impact the potential for a new round of technical development. If a corporate strategy provides low funding for new developments in these areas, this could cause a major downswing in US competitiveness. A scenario can begin with the corporate directors not possessing a strong feeling about funding research areas. In addition, given a scenario of directors not understanding the impact these can cause, the results can almost insure long term product marketing difficulty. This suggests that Congress can anticipate the impact and respond through research at national laboratories. More cross related research is found with regard to transducers.

Gentilman (1996) writes: A description of their new product and test is given. They write: piezo-composite transducers consist of an array of piezoelectric ceramic rods in a polymer matrix. Stiff faceplates are bonded to the composite for stress amplification, when used as a sensor, and to enhance surface response uniformity, when used as an actuator. Many piezo-composite design variations have been produced for specific applications.

Using this process, an entire array of piezoelectric ceramic rods are molded in one operation using specially designed tooling. Several thousand components with excellent piezoelectric properties and part-to-part reproducibility have been manufactured to date. Key words: smart panels, active surface control, piezo-composites, SonoPanel transducers, injection molding, PZT, PZT-polymer composites, accelerometers, and actuators (p. 234).

Piezoelectric devises act something like a switch, producing changes in electrical current when they are mechanically flexed, or the reverse. The development of more complicated, or smart assemblies, drives the need to find computer systems and software, that can be used to control or follow the information presented. The next article below provides an example of responsive change, to meet the new material's technology needs. This has increased the complexity of the technology for R&D.

Ueberschaer (1996) writes: The development of control technology specifically for smart structures and materials has lagged substantially behind that of the base material, transducers, and embedding techniques. Still, development of smart structures with ever-greater numbers of embedded elements continues, spurred by potential uses that require

large arrays of sensors and actuators. A software package is described that offers

improved control systems. A control problem, showing the mathematical model, is

defined. A change of basis for control.

A control problem is discussed. This includes "Multiscale bases—Orthonormal

Wavelets of Compact Support-Compression Properties of

Wavelets". Methods are discussed, for example, "Optimizing Q In Multiscale Basis".
Key

words: smart structures, smart materials, feedback control, robust control, multi-

input/multi-output, wavelets, multirate control This software package, required the

assembly of words from at least six technology areas; Material sciences, smart materials,

control engineering, physical optics, computer sciences, and technical mathematics. One

argues that technical language is changing in our technical society. In addition, the

authors note that their area of research lags behind that of the materials they are

designing software to control. It should be seen that strategic planning is effected by

several aspects found here. These are: The existence of the technology, the diverse and

dispersed research groups, the lag in the software area, the lead of the material, and the

language used. In addition, it is significant that one finds continuity that technology

development and application is not isolated to any one category. This and the previous articles all show overlap and connectivity needs and advantages in science research.

The next author brings in another word-concept. "Intelligent materials". There are language (meaning) differences between those and "Smart materials".

Okano (1996) writes: A research group in Japan offers this report: Considerable research attention has been focused recently on materials which change their structure, and properties in response to external stimuli. "Biomedical Intelligent Systems" are discussed. Authors note that these materials are attractive not only as new sophisticated bio-materials but also for utilization in protein biotechnology, medical diagnosis and advanced site-specific drug delivery systems. (Site specific, for one example would be for an already consumed drug that would only respond to an area that needs the drug as opposed to whole body saturation). Authors show example drawings of a material responding: 1. A hydrophilic versus a hydrophobic response, 2. A material which swells or which can shrink as needed, 3. A material that is either soluble or insoluble or is described as bio-attractive (p. 34).

Confusion can be generated in the differences in the word--concepts for Smart versus Intelligent Materials. There are distinct uses here. A definition and comparison of

Smart versus Intelligent materials is provided next. It demonstrates that quick

conversations about product R&D planning can be somewhat hazardous, unless one is

cognizant of the language differences. A strategy can be centered on development of

smart materials, when the correct strategy was meant to center on intelligent materials.

The difference will effect the actual design features. The new category of research

called Intelligent Materials has evolved from the simultaneous adaptation of multiple

technologies.

The new category has brought about conferences devoted to this area of

research. This is exemplified in a conference report edited by Gobin and Tatibouet,

(1996). An introductory note is provided by the editors. This helps establish the

complexity of the language development around new technology. They write about the

difference of meaning in Smart versus Intelligent materials. Nobody really knows where

the term originated, far less does anyone really know what it means. However, the idea of

the smart structure evolved in the United States in the early 1980s as the potentially

viable solution to the problems posed by the next generation of--particularly military—

aircraft. In parallel, the concept of intelligent materials was pioneered in Japan with the

purpose of establishing a new area in material science to take into account the

interrelations between materials, the natural environment, and society. It was probably in

the late 1980s when the term crept into European vernacular, and over the past decade the

meaning of the words has been discussed, frequently very philosophically and very late at

night, resulting in an accepted interpretation which is quite accurately reflected in the diversity of topic and approaches mirrored in the contents of this conference. (Preface)

It should be seen, from the change in meaning of words sponsored in Japan, that global competition in product development can in fact describe an approach that is differentiated from that found in the United States. One asks how a new category of research can exist globally, if all categories of research are thought to be covered in national research programs.

This suggests that national research perspectives should be broadened to anticipate and respond to the possible knowledge growth that is attainable globally. It is argued that this is a national issue. Another scenario can be defined for this category of language use.

One estimates that research in smart materials could yield results, as spill-over, into intelligent materials. This could cause a need for a quick swing in operations planning. The overall program would need to be evaluated for the issue of whether both categories are the actual subject of focus. This can affect the investment needs. Therefore, a product can be placed on the market that is substantially obsolete, or inadequate, unless intricate planning is undertaken, with regard to these choices, language, subsequent timing, and the basis mission.

In reference to the issue of a definition of knowledge presented by Fusfeld and Haklisch (1994), one argues that a clear definition of knowledge is not yet attained when there is a global difference in research defined as smart versus areas which are categorized as intelligent. One expands the lemma from Fusfeld and Haklisch about a definition of knowledge.

One asks which, or if either of the concepts, smart or intelligent, represents a more advanced knowledge growth and subsequent product development? It is suggested that a holistic national research mission, which creates knowledge, could create elements containing both meanings simultaneously.

89

Tucsnak (1996) writes about control optimization of piezo-electric devices. This provides a cyclical return of the development of material technologies, and the development of computer algorithms to analyze and control them. He writes:

In recent years a lot of papers were devoted to the study of elastic structures with piezoelectric actuators. Concerning the controllability problems, as far as we know, all published works consider finite-dimensional approximations of the initial distributed control problems. The main purpose of the present paper is to study the exact controllability of a beam with piezoelectric actuator by using the theory of infinite-dimensional control systems. Concludes that the exact controllability using infinite-dimensional control systems is an open question (p. 922).

This author, from a research agency in France, explored a new avenue for control. The key issue for this work is that the analysis is about controllability of Piezoelectric actuators. The technology is mentioned in one of the previous articles found in SPIE journals. Therefore, the research and analysis for this intelligent material can be found in at least two other journals and research concentration areas. In other words, it can be argued that a crossover of technologies is continuously growing, and becoming more intertwined.

Research in seemingly diverse areas is found to be connected. Therefore, several managerial questions are created. Is there a need for a research group that is all encompassing? Or, would diverse activities and information supplies be sufficient to insure that a product development is current?

A factor to consider for the answer is that rapid changes can occur as a result from multiple research focus areas, and cause an acceleration of change. It is argued that strategic planning for research on a national scale must recognize and act on the fact that research must be fully aggregated to allow appropriate continuation in developments. In other words, given all of the diverse approaches toward high technology research, rapid changes could be assumed as a potential strategy planning issue. It is suggested that the language for planning in this domain would need careful evaluation.

Information Sciences Dimensions of Research

The information explosion today has become common knowledge. The ability to discover the existence of the complex and interrelated sciences noted above is evidence of that explosion. The fact that in each of the technology areas, researchers are always able to relate to associated knowledge in seemingly diverse areas is further evidence of the knowledge and information explosion. The effects or impacts are referred to as part offuture shock. (Toffler, 1990). National defense strategy includes an ever-increasing quest for improved information. The computer data systems, telecommunications, and CD-ROM systems are part of a network of technology in use today. Information quality and timeliness needs are central to R&D quality. There is no question that the United States can conduct a very large global war. Yet all of the activities depend on correct and timely information.

National policy for Congressional and Presidential action are best served by this as well. The Justice Department has visualized a growing need of correct information, to shorten trial time and aid in crime detection and prevention. Terrorist activities cause a growing need for timely information. (This may become catastrophic in proportion).

Hanson and Day (1994), provide broad information on applied CD-ROM systems used in libraries. Many case studies are noted. Computer network systems are presented in their varied form The text, though current, does not mention the largest of the CD-ROM systems and equipment. In chapter twenty, summaries are given. A Future for CD-ROM as a Strategic Technology ? This chapter mentions data for the amount of storage per CD. This is important when there is no automatic change equipment used. (Many libraries do not have automatic CD-ROM changer equipment)

Advanced automated CD-ROM handling systems are available. These can store and automatically exchange hundreds of CD-ROMs. At the time of a search in 1994, there were three of the systems in use (product has NO NAME for this report). These massive CD-ROM systems were owned by the DOD, the CIA and one other agency— but not LLNL. This does not suggest that LLNL is behind. It may suggest that new methods may be needed to keep up with information which may flow from LLNL.

In addition, the technology found in journals, as shown above, is expanding rapidly. CD-ROM library and search systems are available and becoming more complex. The Society of Photooptical and Instrumentation Engineers (SPIE) journal articles are available on CD-ROM files available over the internet. This is typical of technical journals, and should be known by all practitioners in sciences.

Researchers are becoming more dependent on telecommunications systems. But LLNL reports projects for optical couplers to expand the capacity of telecommunications systems. Therefore, LLNL is found at the leading connection edge in this area.

Section 2 Planning Technology

In section one, technology management concepts were listed. These included comparisons for international planning methods, across-border and cross cultural perspectives. The texts indicate that complex interrelated systems must be considered, and noted that advanced marketing planning is essential for a complete plan. Comparisons were made for various R&D cooperative arrangements.

In this section, planning technology is briefly surveyed in order to argue that overall R&D planning must be intricate. In addition, in this section and section three, literature references show that advanced technology planning and economical development can be viewed as a regional plan.

This perspective is not covered by Summerton (1994). Summerton focused on the product development perspective in acros border, cross cultural, and cross philosophic boundaries, a perspective that didn't account for functional group planning, as in technology regions.

Toffler (1990) suggests that there are considerable social changes occurring because of technology advancements. He argues that these changes affect the entire structure of the economy. Furthermore, independent and isolated activity can become a norm in some social regions, based on Toffler's comments. Examples, as in info-tactics, were given to suggest resistance to fatalistic or top-down decision processes. Therefore, technology regions can be considered as a viable, and perhaps actual choice for some near future planning. Planning for this approach is suggested in the following references.

Friedman (1987) begins by defining the importance of planning and defining planning theory. One could call this a plan for developing plans. He describes planning as a necessity, in all of life, in the market economy (p. 25). He delineates the concept and theory for planning, describing that "the same planning activity may cut across several levels of territorial organization" (p. 25). Some early definitions of elements for plans are given. For this, one includes consideration of the operational definition of planning in the first place. (P. 37)

Friedman (1987, p. 38) suggests that a plan must include goals and objectives, Major alternatives, consequences, evaluation, decision, implementation, and feedback. In other words, one can seek to find if the subject or operation under evaluation really has a plan—and evaluate the plan! Since planning "cuts across boundaries" then the regional planning for a national laboratory environment seems appropriate for a complete evaluation of its activity.

One interprets this text as one approach for elements in a complete planning aggregate. Friedman (1987) teaches the concepts to continually consider planning. His strategy, which includes multiple criteria and interconnectivity of planning, seems to agree with the authors of the earlier texts on technology management.

Planning technology must be included in management technology where advanced large scale research is concerned. Porter (1994) writes: The ideological battles of the past decades between centrally planned and free-market economies, have been settled in favor of the free market economies. Ironically, along with the ascendancy of market ideology, comes the widespread realization that strategic planning, and governmental collaboration, can significantly augment the performance of market economies (p. 2).

Porter (1994) claims a form of paradox. The paradox is that the free market economy plan wins, but governments may help in the new planning. For technology planning, a research management plan is presented.

A Technology Opportunities Analysis (TOA) is explained which is to identify and assess the implications of emerging scientific areas. The process includes analysis of multiple categories which affect technology. Analysis areas are listed : (1) Monitor various literature; (2) analyzing various funding trends; (3) analyzing bibliometric materials; (4) networking with experts; (5) assessing the implications of emerging technologies to present university capabilities, core competencies, gaps, and educational objectives; (6) analyzing policy and action options (p. 2).

The researcher is affiliated with the Technology Policy and Assessment Center at the Georgia Institute of Technology in Atlanta. The items tend to focus on the improvement of the position for research and education for a university based activity. In addition, the region around Atlanta should find local benefit from the advanced information research.

A survey test questionnaire is suggested. He states that a questionnaire should provide a list of technology areas that are thought to be in a high growth pattern. The individuals, who are reported experts in advanced areas, are expected to prioritize the possible developments, based on their expert opinion. This research, in global proportions, is developed for all of the information categories noted above. Therefore, the TO A appears to be strongly based on information acquisition.

Toffler (1990) suggests that info-tactics would become a significant activity of the future. Porter's (1994) article demonstrates a formidable information research activity which has a regional focus. Although one may expect the information to be disseminated quickly, it is argued that this should not be assumed, because when the State of Georgia contributes to the cost of the research, one could expect them to expect return value prior to its free release. In contrast, the information and assessment approach can have failure points, depending on the panel of experts chosen. In other words, a great deal of detailed information is gained, some of which may be inaccurate. One expects a stochastic effect to be applicable here.

Fusfeld and Haklisch (1984) suggested that the university should be central to technology growth, to insure appropriate education quality. They did not mention the formidable plans which this university has demonstrated at keeping up or getting ahead. This would suggest that an entity can make a personal effort to keep up. This also suggests that the top-down approach reported by Summerton (1994) is not the predominant approach found.

Hanson (1996) reports an analysis of research center activities. He studied the agglomeration, and dispersion in a region. He reports impacts on urban planning that are associated with R&D and industrial growth.

He reports that there is a growing tendency to identify regions, rather than nations, as the locus of industrial competitiveness. When considered in this perspective, along with education and knowledge expansion, he argues that the effects are shown to always drive area wages up, along with increases in cost of living. Finally the industries re-locate. This relocation is not immediate upon increase in the costs.

However, when a change begins, it tends to be a large group of changes, and is followed by a new agglomeration. This analysis methodology seems valuable for consideration of the University of California in the San Francisco Bay area. This area can be defined as a region, and the cost of living certainly accelerated in the area. But the Silicone Valley successes contributed to this. However, one estimates that, since there could be a national laboratory agglomeration definition, the normal shift to a new region could be detained, or require new computation methods. New formulation is suggested due to the basis of the computation. Further consideration for a regional comparison could be defined as a factor in the overall strategic laboratory missions.

One would need to consider whether the national laboratory group in that region can be called an agglomeration—by Hanson's (1994) definition. Because a basis for computation was generated using the technology development of garments, which went through a new product phase into production (p. 260).

A product development that is focused can have life cycle effects. Furthermore, the essential plan is cyclical by nature. A question of the direct comparison exists when considering that the national laboratory activities are not the same as a corporate research group. For example: an agglomeration in a region may be driven by speculative investment. Furthermore, business practices are not based on the same financial planning methods or actions as a national laboratory. This is due in part to the effects from Congressional funding.

It is argued that simply comparing a speculative growth and sink for a region, based on business speculation, does not compare mathematically to the university of California consortium, which has been functioning in that region for decades. Although reasonable questions can be compared against Hanson's (1994) agglomeration model, one argues that the method of analysis provides a valuable consideration. Because, increases in cost of living in a region may be affected by a regional plan, despite the form or reason for an investment into research.

Science park development can be compared. Monck (1988) delineates science park cooperative efforts, but argues that this provides efficiency for development planning. However, the external effect could be considered for the regional impact on wages and costs.

One intuitively reasons that, due to the velocity of change throughout the world, a potential for knowledge growth is high. One weighs this against the advantages of the LLNL simply purchasing necessary devices and equipment, rather than attempting to engage in product development, in this world of high velocity changes. But, in direct contrast to this, if the LLNL knowledge base is sufficiently high and prolific, then the marketability of the LLNL technology base should be high, and therefore marketable.

The technology assessment (shown in Chapter 4) gives evidence that the actual productivity of new technology appears to be significantly high. Therefore, the marketability of the LLNL knowledge base should be high. Yet some of this marketability has been hampered by Congressional decisions to reduce funding on technology transfer mechanisms.

Section 3: Government Documents and Analysis

The national laboratory mission and operations are defined by a layer of federal code, regulations and operating procedure guidelines—directives. In addition, patent regulations—law, and contract law affect the operational procedures.

The Lawrence Livermore National Laboratory (LLNL) consortium is operated by the University of California, under contract with the United States Department of Energy (DOE). The consortium consists of three laboratories—one is located in New Mexico, and some operations are in Nevada. Therefore, the state regulations and guidelines for the laboratories include California, New Mexico, and Nevada.

In addition to these controlling boundaries, LLNL engages in functional operations in cooperation with Sandia National Laboratory (SNL), which is located across the road from LLNL, and is individually subject to the same sets of laws. The SNL also has operations in other states. However, SNL is operated by Martin Marietta Corp., under contract with the DOE.

The LLNL consortia, operated by the University of California, does not function under the same set of laws and guidelines as do other national laboratories—as in Argone National Laboratory (ANL) in Illinois. There are cooperative work tasks between LLNL and Oak Ridge National Laboratory (ORNL), which is located in Tennessee. The ORNL is also operated by Martin Marietta Corp., under contract with the DOE. In addition, the contracts in which LLNL is currently engaged include work for the Department of Defense (DOD). The DOD does not function with the same methodology or mission as does LLNL. In addition, the Cooperative Research and Development Agreement (CRADA) contracts which LLNL engages in are with private corporations. These corporations do not function under the same regulations as do national laboratories.

Furthermore, they are subject to regulations by the Internal Revenue Service (IRS), the Securities and Exchange Commission (SEC), and varied state and interstate regulations that are not the same as those for the national laboratories.

Patent regulation and planning is considered. One perspective is provided in the report, Is Today's Science Policy Preparing us for the Future? Opening remarks note that this committee became a full committee twenty years earlier. All government agencies sponsoring research provide documents and comments for this report.

The Department of Commerce (DOC) reports of their Advanced Technology Program (ATP) and R&D expense sharing program. This is operated through their subsidiary agency, the National Institute of Standards and Technology (NIST). They refer to the Manufacturing Extension Partnership (MEP). The ATP provides cash awards to industry for R&D activity. The DOC and NIST report 379 awards to industry and 32 to universities from 1990-1994. The

National Science Foundation (NSF) recorded data that US industry invested $55 billion in R&D in 1993. In addition, federally funded R&D in the same year was $69.7 billion. However, of the federal funds, only 4% were invested in industrial technology, and less than one percent went to the development of early-stage, pre-competitive civilian technologies. The NSF points out that these under-funded activities relate to products 10 years from now. Note that this report gave an outline of the ATP dispersed grants to industry and universities for R&D. This method is contrastive to the government-university-industry plan. It is strictly a cash assistance device that is not centrally located.

The points to be considered from this report are that important projects are under-funded and that there is broad concern, on a national scale, that the United States planning is insufficient. Simply comparing the above books by Fusfeld and Haklisch (1984), Carboni (1988), Summerton (1994), Henry (1991), Gaynor (1996), and Monck (1988), one could conclude that the concerns are valid.

Therefore, an analysis of the LLNL mission is warranted, to discover the ideal research agency for the broad benefit of the US. New technology has driven further questions about how to optimize it. The United States military needs and management have driven expansion in research which is adaptable to a broad range of non- military areas. This is shown in the report, The High Performance Computing and

Communications Program. This High Performance Computer and Communications (HPCC) systems program is not part of the regular funding activities, per se. It is promoted, per this report, essentially by the Department of Defense (DOD). Advantages re given as improvements in early information systems to help prevent unexpected military difficulties. It is given as a means to help reduce the nuclear stock pile while ensuring quality maintenance.

The DOD notes several optimal values for the economy in development of a HPCC program. One includes the Global-Scale Information Infrastructure technologies for improvement of education for the future. Note that this program is not necessarily designed to be located in one central position.

The General Accounting Office (GAO) provides case studies and reports to Congress about other federal agencies. The most notable report provided a core purpose for this project. The report Alternative Futures for the Department of Energy , National Laboratories: The Galvin Report, and National Laboratories Need Clearer Missions and Better Management. This publication also includes "The Galvin Report". The Galvin (1995) report is the report given by this committee headed by Bob Galvin. The

Galvin committee was a special committee, solicited by some agency, to provide an opinion from a non government group. Bob Galvin is noted as the Chief Executive Officer of Motorola Corporation. There are key elements to both the GAO report and the Galvin report. The GAO testified to their findings. The DOE and national laboratory officials offered evidence to the report. In addition, the GAO offers commentary which is in disagreement to items reported by the DOE.

This 1240 page evidentiary document provides general data, testimony and comments by the laboratory directors, GAO accounting, DOE defense claims, and Galvin (1995) committee findings. There are key claims by all parties that argue the title of the report—that the national laboratories need clearer missions and better management.

- ♦ GAO reports that the DOE and national laboratories do not manage their sub contracts well-that severe problems and waste exist (p. 224).

- ♦ GAO reports that the DOE does not manage the laboratory activities well (p. 225).

◆ Galvin committee reports that the DOE has overburdened the laboratories with micro- management and points to management policy difficulties that 'rest at the feet of the DOE and Congress' (p. 32).

◆ Galvin committee reports that the contract structure for operating the laboratories should change to more resemble industry. This is offered in two forms in response to their findings that the laboratory functional operations are not managed well (p. 39).

◆ Galvin committee reports that technology development is not effectively beneficial to industry and offers that a new structure be implemented (p. 18).

◆ Laboratory officials agree that improvements are needed, but that complete changes in contract structures are not needed. Officials claim some of the difficulties noted by the GAO, are correct (pp. 101-112).

◆ DOE claims to have reduced the number of project management personnel and restructured other guidance methods (pp. 77-321).

◆ Laboratory officials disagree that there is little effort in technology transfer (pp. 115-117).

◆ LLNL director claims that the central mission of this lab is nuclear weapons and stockpile maintenance (p. 130).

More perspectives for the needs to analyze national research programs is found in the report, Restructuring the Federal Scientific Establishment; Dismantling of the Department of Commerce.. The multiple subsidiary agencies and overall effectiveness are considered. Reports show that several agencies are involved in overlapping and interactive R&D programs. The NIST reports again of their ATP and MEP programs. In addition, The National Science Foundation (NSF) reported that they have 1500 principle investigators involved in support of ocean sciences research. The NRC reports that there are a collection of more than 2100 researchers involved in this one topic-when cross comparing 12 government agencies. One agency listed is the DOE.

Although the only agency that mentioned sponsoring industry R&D was the NIST, it would seem obvious that all of the other research programs need equipment from industry . In these research activities, markets are created for industry that then can afford to be involved in research. The hearings note reports from multiple agencies. However, given all of the management objectives represented, the report was not a complete representation of needs and assessments for research activities.

National planning for education of scientists is a cross issue under consideration by Congress. The report, Reshaping the Graduate Education of Scientists and Engineers: NAS's Committee on Science Engineering , and Public Policy Report, notes difficulties and recommendations for improvement of graduate education for scientists and engineers.

Recent PhD. graduates are unemployed for longer periods of time following

graduation, and is noted that there is inefficiency in preparing PhD's for employment. The

report recommends a new class of grants, partly to broaden the education of the Ph.D.

It is further recommended that a national education data base be formed. One questions

the rationale of this report. Of all of the PhD's involved in federally funded research, and

of all the PhD's employed in federal agencies, there seems to be no automaticity to

affective Ph.D. education planning. One asks who is really the authority on this process

or need? This suggests the need to examine the basis for education and planning. Along

with the issues of planning and budgeting, law issues become prominent. The report

Laws Relating To Federal Procurement covers military procurement details. It also has

provisions for withdrawal from a prime contract in a region. This notes that if there were

a withdrawal from a statistical region with an impact toward 1000 or more jobs with a

certain skill area, that the DOD is authorized to provide financial compensation to the

impacted state or region. In other words, the DOD has the authority to make the decisions

to move a large-scale weapons development or procurement operation, but it must

pay some compensation. In other provisions, the DOD is instructed to not distribute

prime contract purchases and development in a way that will not be unfair to other

national regional development.

One intuitively reasons that this boundary can effect the DOD planning that may not produce the most ideal product. Because the boundary does not depend on the DOD need, it is focused on money distribution. One considers this combined regulation group, as the concept to not cause accelerated regional development in one area, at the expense of another. In consideration of the analysis by Hanson (1996) noted for agglomeration phases, these DOD federal directives are devices that can reduce the severity of impact, or cause major changes in a region.

The budgeting and planning for national research finds issues with regard to technology transfer. This is demonstrated in the report, Federal Technology Transfer Policies and our Federal Laboratories: Methods for Improving Incentives for the Technology Transfer at Federal Laboratories: Joint Hearing, Before the Subcommittee on Technology, and the Sub Committee on Basic Research.

This testimony and summary of meetings provides notation of codes that include Cooperative Research and Development Agreement (CRADA) contract guidelines. One key item is under Technology and innovation—Chapter 63, United States Code Annotated Title 15, Commerce & Trade, Sections 3701- 3715 (as amended through 1993 public laws and with annotations) (p. 21). Considerable debate developed over the CRADA provisions with relation to this and other codes.At one point, the CRADA regulation omits the congressional code, title 15, section 3701.

At another cross-road, the consideration for the employees' rights under Government Owned-Contractor Operated (GOCO) are at issue. In another debate, the point is made of a CRADA participant needing to petition for a license to use a patent developed under the CRADA arrangement. It was argued that the participant was required to petition the DOE for each occurrence in spite of having a CRADA contract during its development (p. 139).

This continuous mix of operational conditions for innovation development will hamper innovation. It seems significant that part of the debate developed in methods to get around an existing law. Furthermore, the entire hearing premise is to encourage the laboratories to be more effective in their business dealings and R&D operations. The conditions of this evidence found in hearings can be weighed for the effect on the velocity of innovation as noted earlier. In addition, one of the key issues is product liability. The laboratories do not actually want to give up a patent. However, Congress refuses to share in the product liability (p. 128). One other key issue found in the same group of conflicts is the provisions in a CRADA contract which allows the laboratory March in Rights —to be able to walk into their business partner's premises at any time to inspect or inquire.

Long (1995) reports that technology transfer has had extremely poor success. In one category of application, absolute failure occurred in environmental security. Long (1995) provides a detailed report about failures in environmental technology transfer. The original hypothesis test showed absolute failure. In fact, the technology transferred created large-scale environmental damage. Long writes that the processes used for technology laboratory and cooperative University-Industry productivity.

He provides facts which show that the number of patents generated in research is far greater than those that actually find their way into marketable products. He shows a comparison of R&D expenditure, compared to the actual product use of patents, results in a cost of $571.4 million dollars per patent (p. 5). This is because, in spite of the many patents claimed by the government, only a few are developed into products.

Long's (1995) findings may be compared to the previous report on education planning. The Congressional hearing contributors provide evidence that planning must be improved. Long's findings show that failures do occur that are due to insufficient planning. It may be concluded that education and development planning require improvements. One key element that is apparent in both reports is the omission of important analysis planning elements. In other words, broadened analysis methods are shown to be needed at all levels of planning.

In examining the issues of technology transfer, one must include analysis of strategies that inventing agencies must use to protect their innovation rights. Thus, complex issues arise in all perspectives of research on a national and global scale. Tager and Von Witzlegben (1990) report on a European community conference on the topic of innovation protection. The conference was the first European congress to deal with patent and innovation matters. It focused on how to promote innovation, and accelerate the spread of new technologies, by means of the patent system. Thirty leading European community experts contributed—this includes the interaction of patent attorneys, and other patent consultants. Some are quoted here, and referred to by the country represented. Spain: The problem is to cut the cost of innovation and distribute information of innovations" (p. 8).He reports that they conduct technological surveys and distribute the information as quickly as possible.

Economic Community directorate: Director-general of Telecommunications, information industries and Innovation —Commission of the European Communities. The overall European community plan for innovation exploitation includes all of the countries. They note one of the problem areas as being the industrial orientation programs that are going on in the European community (p. 11).

Germany: The Implications of Patent rights for Corporate Innovation Policies in the Future European Single Market (p. 17). Firstly, patenting is to be given pride of place in innovation and general economic policy. Secondly, emphasis is to be placed on the importance of the protection and information value of patenting for competition and innovation policy. The innovation and marketing experts concerned are to be made more aware of this. Thirdly, technology transfer experts are to be helped in the difficult task of liaising between companies and research establishments.

Denmark: Promotion of Innovation: A new direction for national patent offices (p. 39). About competition: The realization of the single market as from 1992 will change the competitive situation considerably. The competition will be more intense — everywhere— also on the markets which we today call home markets. The situation that, for example, Danish enterprises should occupy a place apart on the Danish market, is becoming history.

France: Public Research Institutes: Beyond Academics-Innovation as a new task (p. 97). There is no doubt that the answers to the major questions facing our society are to be found in scientific knowledge, in the widest sense of the term. Research organizations therefore occupy a particularly important place in French society as the driving force for economic development and source of progress for mankind.

France: Cost and Risks of Patent Infringement Litigation Possibilities for Conciliation (p. 179). Now, according to some statistics, out of 60 new product concepts, only one sees the light of day....according to the chairman of American company FMC, 'to sell a single new agri-chemical product, we have to test 15,000 in the laboratory, i nvest $60 million and probably half the life of the patent will expire'.

This European Community conference included participant speakers from more than 15 countries. Of the articles noted here, significant cooperative efforts are shown, as well as frank understandings of the issues of patent and innovation risks and adventures. The speaker from Denmark reported the depth and speed of changes make the competition within the parent country difficult. The speaker from Spain reported their efforts to disseminate innovation information as quickly as possible. But, the speaker from Germany indicates that innovators should be given priority and assistance for technology transfer.

The collection of reports indicates that the issues about innovation, and the fact of coming change, are well understood in the European community. This suggests that the US policy should not demonstrate hesitancy, or mixed strategy to maintain US economic security, and international cooperation. Returning to the issue of law matters, one must examine the perspective of the actual legal issues within a complex research agency.

Gilboy (1995) considers, in detail, the claims of effects on a government agency caused by outside pressures. She notes the classic attitudes are: Accommodation; of external pressures, such as those caused by interest groups. Amplification: where officials enlarge the effects of external influences by anticipating the potential likely consequences of certain kinds of agency actions, based upon the understandings of how situations have affected past developments. Assimilation: Where agency officials define a situation as problematic and coordinate with, or take into account, various elements of their environment in attending to the perceived problem.

Gilboy (1995) reports studies in several states yield varied results about the meanings of accommodation from the agency. For example: "Agency administrators interviewed in Minnesota and Kentucky reported that 23 to 44 percent of the time legislator's requests on behalf of constituents amounted to 'special favors' (p. 7). This suggests that the three predominant discretionary powers noted above can be driven by changing environments collectively caused by legislators. Gilboy suggests that generalizations will be difficult without contextual sensitive approaches to understand behavioral irregularities (p. 5). This article, which considers the forces affecting government agencies, is from

Law and Policy journal.

Section 4: Evaluation and Management Methods.

Operations Analysis of Engineering Sciences, the mission of Lawrence Livermore National Laboratory (LLNL), as the title of this research project, was undertaken as a multiple layered, and multiple topic Meta-Analysis. Collected force systems (vectorial in character) are envisioned as having simultaneous effect on the operational efficiency and the overall mission. In addition to the external forces affecting the Engineering Sciences Operations and mission, the actual internal knowledge production processes in the development of research programs effects the final mission. Therefore this operations analysis is focused on the LLNL Engineering Sciences Operations, and its mission, for achieving the LLNL mission.

The Case Study analysis for this study uses classical evaluation models, primarily drawn from the methods defined by the General Accounting Office (GAO). But this methodology is also aggregated with other evaluation methods and devices. These methods offer significant tools for discovery of information that are not necessarily mathematical formulations. However, as is shown in the analysis from the research using these methods, mathematical formulation is a primary core skill area discovered at LLNL. In contrast to review and performance evaluations, mathematical modeling for operations exist which have a proven effect on performance analysis. Furthermore, some classical models for interview methods can be extrapolated into strict mathematical modeling.

Curiously, for this to be effective, interview methods that employ language, and not math, often provide the starting means for the evaluation. This point is delineated in the analysis of the research which is shown in Chapter 4. But the GAO methods, as well as those of Lacy (1992) shown below, and those of Saaty and Alexander (1989) can begin by interviews.

The references that follow include current mathematically oriented tools as well as interview oriented methods. The DOE and the GAO have evaluation methods. These are included as reference since they have all been used for current evaluation means. It is important to note that the DOE testified that they have and use their published evaluation methods. In contrast to this claim, the General

Accounting Office (GAO) used its evaluations methods to show that the LLNL is not managed well by the DOE. Federal law which states that the General Accounting Office (GAO) is the only federal agency created for the sole purpose of evaluations. Their

operations center on this. To achieve an analysis of the LLNL mission, new avenues of analysis about research are explored.

Therefore, consideration of the operational scope or dimensions for the LLNL engineering sciences operations research and development (R&D) capabilities are warranted, in order to understand its true mission.

The primary research activities toward knowledge production are seen as fundamentally important for this project. But none of the performance or evaluations methods found actually examine the principles of knowledge production. In addition, the Congressional hearing documents demonstrate that significant controversy exists with regard to methods of accounting. Therefore, performance evaluations, which ultimately track or are focused on accounting, do not define the final measure for value to the United States, because they do not analyze knowledge production.

One recalls the issues of measure, defined by Von Neumann and Morganstem (1947/1972), that the basis of measure is where the controversy is found. In addition, Fusfeld and Haklisch (1984) argue that controversy exists with regard to a definition of knowledge. In addition, authors continually agree that the prediction of innovation is not well defined. See Henry (1991), Gaynor (1996), and Carboni (1990).

One asks how the DOE believes that a performance measure provides accurate information. One expects that the DOE is using its data from its performance evaluations. But the GAO has argued that the DOE has not managed the national laboratories well. In spite of all of those issues, the LLNL has produced many products, as shown in the Technology Assessment of this project.

The DOD uses mathematical game theory for operations and mission planning. These are shown to be applicable to experimentally examine the LLNL operations. But a Research program would be needed to experimentally examine the many possibilities. The Congressional hearing reports show controversy about the laboratory missions and about the DOE management of the missions. Therefore, this researcher examined the methods used by the GAO in detail. In addition, the GAO offices in Washington DC were

visited.

The GAO internet web page was searched. All non classified reports that are generated by the GAO are obtainable by direct order on the GAO web page and order screen. Most of the regulations which the GAO have designed for their government accounting methods can be ordered for two dollars. The question that was in mind centered on how the GAO would be able to make a report to congress with the title National Laboratories Need Clearer Missions and Better Management: April 12, 1995.

One method of gathering data for a case study is by means of questioning a group of well known experts. The GAO designed a specific questionnaire for this purpose, and reportedly followed this by holding a conference. The questionnaire followed the concepts and processes in the regulations noted below. The conference allowed the experts to share their thoughts openly, thereby drawing further from their mutual understanding about the missions of the laboratories.

The experts were apparently collection of individuals who had some experience directing a large federal research agency or group, but who were not connected with the DOE at the time. Therefore, the report from the GAO was based specifically on collected comments from experts. It provided valuable information for Congress, which is shown in the reports and hearings. However, this did not end the controversy about the missions of the DOE and the laboratories. Nor did it measure the issues about knowledge production and innovation prediction. Therefore, more details are examined below as to the techniques of the GAO for gathering data.

These are supported (in Chapter 4) by specific analysis of knowledge production and subsequent innovation. But the basis data gathering concepts used are from the regulations that are shown below. In addition, ideas from other authors are surveyed below.

In the GAO regulation, Designing Evaluations, the GAO advises that many diverse methodologies are needed to develop sound and timely answers to the questions that are posed by Congress. This report teaches fundamental considerations that may be required. It advises that appropriate negotiations are needed to ensure that the information gathered in an interview will net an affective work. (p. 27)

Designing Evaluations, GAO/PEMD-IO. 1.4, is the GAO report called DE 1.4 for this project. The GAO also provides regulations titled Case Study Evaluations (GAO/PEMD-IO.1.9), and The Evaluation Synthesis, GAO/PEMD- 10.1.2. The concepts and methodology in the regulations directly interrelate. The DE 1.4 explains:

♦ But what exactly is an evaluation design? Because there may be different views about the answer to this question, it is well to state what is understood in this paper. Evaluation pertains to the systematic examination of events or conditions that have (or are presumed to have) occurred at an earlier time or that are unfolding as the evaluation takes place. But to be examined, these events or conditions must exist, must be describable, must have occurred or be occurring. Evaluation is, thus, retrospective in that the emphasis is on what has been or is being observed, not on what is likely to happen (as in forecasting) (p. 6).

♦ To further characterize evaluation design, it is useful to look closely at the questions we pose and the answers we seek. Evaluation questions can be divided into three kinds: descriptive questions, normative questions, and impact (cause-and-effect) questions. The answers to descriptive questions provide, as the name implies, descriptive information about specific conditions or events. ...The answers to normative questions (which, unlike descriptive questions, focus on what should be rather than what is) compare an observed outcome to an expected level of performance. The answers to impact (cause-and-effect) questions help reveal whether observed conditions or events can be attributed to program operations (DE 1.4, p. 7).

Given these questions, what elements of a design should be specified before the information is collected"? The most important element can be listed as:

1. Kind of information to be acquired.
2. Sources of information.
3. Methods to be used for sampling sources.
4. Methods of collecting information (for example, structured interviews and self- administered. questionnaires).
5. Timing and frequency of information collected.
6. Basis for comparing outcomes with and without a program analysis plan (DE 1.4, p. 7).

The DE 1.4 provides other detail with regard to the foregoing list. It continues with comparisons of varied levels of evaluation. This pertains partly to the cost of making an evaluation, and the available material. These are referred to as a "Strong or a Pilot evaluation" (p. 15). Once again the concepts are relevant to a case study or a survey research method. (p. 19).

The GAO noted that a cost issue can affect the available interview subjects. For example: Where directors of the DOE have been evaluated, the cost of their time and travel arrangements could be considered. A rule of thumb is offered: The time for a study

and the scope of the question being addressed ought to be directly related. Tightly structured, and narrow investigations are more appropriate when time is short. Any increase in the scope of a study, should be accompanied by a commensurate increase in the amount of time that is available for it. The failure to recognize and plan for this link, between time and scope, is the Achilles heel of evaluation (p. 19).

The DE 1.4 provides a list of evaluation types.

> Sample Survey
> Case Study
> M Field experiment W
> Use of available data (p. 23).

The preceding definition provides a broad means to explore a case. In other words, prior case can be reevaluated for new evidence. In addition, the concept is usable to discover what evidence is not available. Furthermore, one could use the concept to define where prior studies should have included more detail.

The GAO advises that not all evaluations can be designed around mathematical formulations. Nevertheless, these can be effective, under staff expertise and cost issues (p. 21). The GAO/PEMD10.1.9 (DE 1.9) provides general concepts for definition of Case Study. It is shown that the concepts allow broad, or flexible use. The DE 1.9 advises:

> ♦ The case study strategy is less well defined than the other evaluation strategies we have identified and, indeed, different practitioners may use the term to mean quite different things. For GAO's purposes, a case study is an analytic description of an event, a process, an institution, or a program (p. 40).

The GAO regulation, The Evaluation Synthesis, noted as GAO-PEMD 10.1.2, is referred to this paper as the ES 1.2. This regulation provides materials for constructing a study of complex issues and agencies. In addition to the title, the concept of development of an Evaluation Aggregate is presented. The ES 1.2 advises:

> ♦ Five types of information are potentially valuable for the evaluation synthesis that are not suitable for statistical analysis are (1) single case designs, (2) non-quantitative aggregate studies, (3) non-quantitative information in quantitative studies, (4) expect judgments, and (5) narrative reviews of collections of research studies (p. 48).

♦ Expert Judgment: An evaluator may choose to include expert opinion at an early stage of the synthesis, such as in evaluating individual studies. Instead, the evaluator may want to systematically compare studies relying on expert judgments about program effectiveness. Synthesis should be able to incorporate these inputs (p.51).

♦ Under special circumstances, non-quantitative approaches to the evaluation synthesis
are particularly appropriate. For when, (1) treatments may be individual or more concerned with process than outcomes, (2) program effects are assessed across multiple levels of effect, (3) uncontrolled treatment groups are compared with the treated control group, and (4) the "wrong" treatment is studied (p.51).

In essence, the issue is to find a means to provide useful comparative evidence that make use of available information. In most instances, the synthesis and aggregate are to help provide insight for complicated issues. In addition, these forms of evaluation can provide some form of rationale for possible policy issues.

The official GAO regulation handbook, Government Auditing Standards: GAO-Yellow book: provides a broad perspective of needs for analysis of government agencies. This 110-page regulation provides the primary operating rules and procedures used by the GAO. This is supported by multiple categorical rules and regulations that the book lists.

Distinctive methodology is defined for financial audits and performance evaluations. Key factors are noted for performance audits. These include evaluation of the material evidence quality (p. 86).

The references can be compared to the DOE performance and systems analysis

methods. However there was DOE failure in key areas, according to the GAO. In other

words, one must determine whether the LLNL operations have any consistently

measurable activities when considering the effects toward the final mission. An alternate

evaluation method is found in the following document. The DOE apparently uses this

book, or components to ascertain measures of performance.

In contrast to the evaluation methods defined by the GAO, the DOE claims to use

an internal evaluation handbook. How to Measure Performance; Handbook of
Techniques

and Tools.(TRADE) (1995); Prepared by the Performance Based Management Special

Interest Group. Department of Energy (DOE). This book begins with a description of a

large group of contributors. It also begins with a very lengthy legal disclaimer notice.

Both are noted here, since this methodology is claimed to be used by the DOE.

Prepared by the Performance Based Management Special Interest Group.-For the

Special Project Group Assistant Secretary for Defense Programs and the Office of

Operating Experience, Analysis and Feedback, Assistant Secretary for the Environment,

Safety and Health, DOE.

The Training Resources and Data Exchange (TRADE) network is managed by the Oak Ridge Institute for Science and Education (ORISE). This is part of the Oak Ridge Association of Universities (ORAU) which is an association of 88 universities that was started in 1946. None of the participants (listed above-and the printers and DOE employees) assumes liability or responsibility for the accuracy, completeness, or usefulness of any information, apparatus, product, or process disclosed, or represents that its use would not infringe on privately owned rights.

The DOE claims this document and its technology as fundamental to their evaluation practices. Yet the GAO was able to say that the DOE and the LLNL are not managed well (the GAO and Galvin rested LLNL management failures on the DOE). This DOE document shows a careful, perhaps overstated consideration of liabilities. On the other hand, the document promotes network systems for personnel needs. One asks if the issues of obtaining employees by means of a network was considered for its impact or conflict with federal statutes on employment opportunities?

In other words, the overall DOE operations include the interacting forces of law and politics. Issues of laboratory operations and mission could be impacted by a DOE law/politics force factor. This should be considered with regard to the potential university cooperative funding, which includes many universities that cooperate with the National

125

Science Foundation and the Department of Commerce.

Given the contributors listed, the book does not articulate a means for precise measure of knowledge creation and innovation production. One must consider non-government evaluation methods. Sichel (1982) provides "program evaluation guidelines." She provides heuristic concepts for considering and designing an evaluation.

Sichel (1982) begins with a precautionary note about data from evaluations. To Satisfy Powers-That-Be! In the real world, substantive questions may be secondary to the pushes and pulls of organizational hierarchies, interagency relationships, and governmental bureaucracy, in getting evaluations started. There may be strong political pressure for program evaluation—the media may be demanding accountability from. Evaluation can conclude that programs are either unnecessary or uneconomical, or working poorly for some other reason. Some motives of Powers- That-Be are open and admitted; some are not, as with personal aggrandizement concerns and other organizational infighting" (p. 13).

In chapter two, further heuristic considerations are given. "Information Systems vs. Informative Answers, .and BUT WE ALREADY KEEP STATISTICS. (p. 15). These headings and subsequent concepts suggest that Informative Answers may be more accurate than many files of statistical data and analysis.

125

Zey-Ferrell (1979) edits an evaluation book which features consideration of the "dimensions of organizations." She provides key concepts and measure limits for evaluating organizations. The chapter topics and study include key words, such as: Environmental Dimensions, Core Typology, Contextual Dimensions, Structure and Process Dimensions, Strategies of Control, Performance Dimensions, Systems Resource, and finally the Measurement of Organizational Effectiveness.

Concepts are shown for finding and understanding the dimensions of organizations for subsequent evaluations. Key sub topics define Limits of Strategic Choice, and other similar measurement methods. The text does not spiral into theoretical mathematical simulation modeling. It does help define organizational efficiency through heuristic boundary definitions, called dimensions.

It should be seen that Zey-Ferrell's (1979) method is to find dimensions of an organization. But, when considering effective research organization, this researcher is seeking dimensions in a broader frame of reference and connected patterns. It is shown in Chapter 4 of this project that the issues of connected concepts and dimensions have a more detailed connection than is presented by simply defining numerical dimensions for the management. Lacy (1992) provides concepts and techniques for developing an operational systems evaluation.

This is afforded by the author's plan to cover a wide area of potential adaptation. This can be a computer systems evaluation text. However, it presents the general concepts for evaluation of an operation. It provides heuristic methodology and concepts to assist in data collection. This is largely done through the interview of the actors. It can provide the concepts needed to collect data in the form that can be interpolated into a mathematical model or a heuristic generalization.

A sequence of analysis parameter characterization is given. The information search is given in the sequence topic areas, and then broken into subcategories in the following overall format:

1. Problem Definition: To examine the issues, or problems that create management difficulties.

2. Value System Design: To examine the key reasons and relationships for the existence of the operation.

3. Function Analysis: To examine the real activities within the operation.

4. Systems Synthesis: To construct and test an operational synthesis, based on the discoveries of items 1-3.

5. System Analysis: To examine the operation, using any possible mathematical models that the operation or its synthesis seem to show.

6. Decomposition: To turn the new analysis and synthesis discoveries into a revised operation, where applicable.

7. Description: To define the actual operation or mission, based on the discoveries and evaluation of the evidence.

Methodology is provided that may be adaptable to any functional operation. Therefore, the tools provide methods to evaluate complicated topics, as well as overlapping considerations using the same methodology. A key element for this text is the consideration of all possible attributes that effect an operation.

Lacy (1992) makes the point that if there is an incomplete plan, then the quality will be reduced, and the risk will be increased. The needs for more complete plans in CRADA application processes are shown as fundamental need by Long (1995). He reported that technology transfer through CRADA agreements, generally and historically do not work well. Therefore, an improved analysis of the operations and the mission would suggest that improvements can be made.

Standard management analysis methods must be considered when analyzing the complex research agency. Although this approach is used for this project, it is shown that the current methods do not provide adequate means to examine the complex perspectives of complex research on a national scale. The issues are considered above with regard to the complex mathematical foundation that is shown in human action.

The text by Lee (1987) is one of several on the topic of management sciences that are used in this project. Linear programming is presented with great detail by Lee (1987), along with comparative probability trees. Models and graphs are shown for development of concepts of queue theory and applications. Regression analysis and trend evaluation are shown. Therefore, linear programming is the primary analysis tool for this text. It is shown that the analysis for complex research demonstrates non- linear formulation, as well as linear differential needs. See Chapter 4, and Basare and Haurei (1994),

Dresher (1981), and Kloeden and Platen (1992). Although Lee (1987) gives detail f for evaluation of the measurable as aspects of a system, there is no clear notation of failure risk when part of the analysis material is omitted. This text defines a great deal of useful mathematical analysis tools. However, in addition to not delineating failure potential, it is not a strategic management text. The Lee (1987) text is generally compared, but not delineated in the analysis of Chapter 4. Concepts from this and many other texts must be aggregated to formulate a more complete analysis. One text that includs strategy analysis is described below.

Thompson and Strickland (1987) provides overall programmatic analysis concepts and methodology for corporations. The text analysis methods can be compared to the development of a laboratory mission, since it includes evaluation of the balance in a strategic plan. Considerations are given for divestiture or acquisition planning of small and major product development areas.

A distinction should be considered, that some of the activities found at a national laboratory are not necessarily balanced, due to input from varied sources. In addition, this may be affected from the actions of the DOE and the issues that are considered in the GAO reports. However, the basic principles covered here provide useful concepts for strategic planning including balance, divestiture and acquisition. This does not spiral into an advanced mathematical analysis text.

The authors delineate the importance of having a balanced strategy— one that does not include centric elements that can detract from the overall corporate strategy. This suggests that even a program that shows a phenomenal growth may only be temporary, but detract from the original strategic plan. Stages of development must be considered (p. 218). Geographical logistical considerations are suggested.

This element of analysis can be with regard to the physical as well as cultural issues. Analysis of trends are suggested for strategy planning (p. 60). The analysis of trends are with regard to potential marketing of the products. The trend issues may also be with regard to potential costs and overall corporate impacts. An important advantage of this text is that it covers the ideas of strategic analysis. A great deal of calculating could be accomplished that will produce a failure if there were important issues omitted. Therefore, it is argued that a strategic analysis of an entity which is primarily research centered, should include appropriate computation and evaluation methods, with regard to knowledge production.

CHAPTER 3

RESEARCH METHODOLOGY

Analysis of the Lawrence Livermore National Laboratory (LLNL) mission and engineering sciences research operations was accomplished using a layered meta-analysis. The analysis layers provide triangular validation for a more clear understanding of the LLNL mission. This is defined as the research and survey of reports and evaluations that are in existence and have been accomplished by government agencies, including reports of projects accomplished by LLNL. In addition, aggregation of the data provides cross or triangular validation.

Cross comparison of documents from hearings about other national research activities were collected for comparative analysis; this included federal statutes. In addition, regional planning for the San Francisco bay area was considered from California and local planning authorities. The primary data on regional effects is found in LLNL publications.

The analysis objectives were to find significant information about the LLNL engineering sciences operations. The operations methods impact the overall effectiveness of this entity. A significant effort was made to ascertain the core values of the operations and systems that may function toward the national good by means of the actualization of the answers to a set of questions, listed below, were sought that are noted under the heading Systems and Operations Questions.

The answers to these are cross compared to standard operations practices that can be found in current text and scholarly journals. The list of questions is written following the definition of the evaluation aggregate. The answers to the questions are given in Chapter 5. Although the word evaluate is drawn from the methodology provided by the GAO, this project is more exactly an analysis. This is an analysis of the LLNL mission. This includes study of the theoretical principles of knowledge production and subsequent innovation production. Significant evidence was discovered which suggests that a primary difficulty facing Congress is that there is no clear definition of a mission with regard to theoretical knowledge production. Therefore, this project was defined to include that appropriate criterion.

The Analysis Aggregate

The components for the evaluation of the LLNL mission are an aggregate. They are constructed using evaluation technology, defined in the literature review of Chapter 2. Since evaluation methods have been developed by the GAO and reference is made to its technology, key terminology is drawn from its texts on evaluations.

The GAO document, The Evaluation Synthesis GAO/PEMD-IO.1.2, provides concepts for collecting prior evaluations. All prior material is gathered, re-evaluated, and supported by any appropriate new material that can give new understanding. This method is referred to as a construct—an aggregate. The new construct is referred to as the Evaluation Synthesis. For this project, the aggregate method is considered fundamental.

The aggregate construct for this project includes multiple layers and categories of methodology. The LLNL engineering sciences research operation and subsequent laboratory mission represent both complex system, and one that is operational in an extremely broad environment. Therefore, the analysis components were implemented in consideration of the complex criteria. The analysis collection must be considered as experimental. It should be intuitively understood that, although each methodology may not present a complete analysis, the aggregate provides a triangular validation for understanding the operational characteristics of the LLNL toward actualization of its

Each category of analysis listed below must be considered as an experimental device and element of the whole. However, although a systemic data analysis is presented as one of the elements, the overall methodology is considered as a search and analysis using systemic concepts. The systemic principles for this program are methods t to define a construct of the aggregate for the analysis synthesis.

There are seven components that are used for the aggregate construct. In the research activity it was discovered that the components listed next overlap and are connected. Each of the topics listed below are part of the analysis, but area headings are grouped in the analysis of Chapter 4. The research area reference list is as follows:

1. General Operations Analysis.

2. Strategic Management Data Composite.

3. Systemic Structural Analysis.

4. Technology Assessment.

5. Information Sciences Strategic Plan.

6. Operations Analysis of Engineering Sciences.

7. Technology Region Survey.

Analysis Aggregate Component Brief

General Operations Analysis

This includes data about the legal foundation of LLNL and the conflicts that are found in the Congressional hearings. The overall LLNL operations are found to be impacted by the DOE and Congress.

The hearings revealed that LLNL had difficulty, and experienced inefficiency attempting to follow all of the DOE directives. (See the 1995 Galvin report). Therefore, this component discovery was that LLNL has capabilities that are not optimized, because there are boundaries established by the DOE and Congress. For example: LLNL has funding to accomplish a group of research programs, but are restricted from direct product development. The issue of advanced optimization of R&D potential is delineated in Chapter 4.

Strategic Management Data Composite

Text methodology was aggregated and cross compared in a composite of the Systems and Operations analysis components. The text by Thompson and Strickland (1987) was used for comparisons of the strategic business posture analysis. This text defines methodology for strategic planning for business enterprises. Chart methods, which consider the strategic operations, were drawn to define the aggregate of a strategic operations plan and analysis. These were supplied by the LLNL Director. A new chart was drawn as an interpretation of the LLNL charts. This is titled the Strategic Operations Plan, and is examined in Chapter 4.

Following the key asset evaluations, a marketing analysis was undertaken. This is followed by analysis of key methods for strategy. A few of these were given as product differentiation and optimization. In other words, the key strengths analysis shows that there are areas of development which can be optimized further. Discussion of the product differentiation issues and knowledge production principles at LLNL are compared in a general discussion of other issues that consider mathematical analysis.

Thompson and Strickland (1987) offer strategy considerations such as creating a larger product differentiation and thereby improving the overall business profit strength. Other methods are shown for consideration of acquisition and divestiture. They consider divestiture for the emotional issues that are found in a corporation. Large aggregate corporate activity is shown for the effects on cash and credit issues. These in turn are considered with divestiture and acquisition. One finds important elements, as in key corporate technology that can act as a serious detriment to a broad corporate strategy. However, their methods do not spiral into the complex computation systems noted above.

The overall data elements for this area of analysis of LLNL are shown to be comparable to the strategic management concept shown by Thompson and Strickland (1987). However, one can define a nuclear weapon driven technology as a plus or minus on a central strategy. Therefore, other methods of evaluation were articulated for an aggregate strategy comparison.

The text methodology by Lee (1987) was cross compared. This text defines model development that is based on measurable data collected in an evaluation. A systems methodology is claimed (p. 7), which is given as a systematic approach to decision making. A data collection phase is defined. This includes assembly of data in measurable relationships. Lee provides model classifications for both mathematical formulation and communication (p. 21). Linear programming is articulated by Lee to help optimize measurable data.

It is shown in Chapter 4 that, although the text models referenced above for this area provide valuable information, they do not offer the complex computation systems analysis that is demonstrated at LLNL. Specifically, the texts and the software utilize linear programming or linear computation methods, but it is demonstrated that the LLNL complex research requires both linear and nonlinear stochastic partial differential computation needs. This is suggested in the theoretical mathematics by Kosco (1992), Basar and Haurei (1994), Kloeden and Platen (1992), Churchland and Sejnowski (1992), Dresher (1981), Henry (1991), Gen and Cheng (1996), that are noted above, and Eisburg and Resnick (1974), noted below.

Systemic Structural Analysis

A systems analysis, which develops data structures, was used to help define the complex needs assessment for LLNL. The text methodology was used from Lacy (1992). This text provides a systemic approach to develop a needs assessment. They note that a singular activity, as in one key research area, can be evaluated for the complex interdepartmental, organization, financial, and strategic objectives.

The data collection and structuring methods shown were used to help discover the LLNL aggregate that is defined by the GAO. One defines this method as use of an expert shopping list to define important elements which affect an organizational plan. Lacy (1992) provides a group of categorical analysis lists to discover systemicy (shown in the literature review)

Lacy's (1992) methodology is used as a key device for primary definition of key elements found in the LLNL research operations. The Lacy list method helped discover data and structuration tools for chartforms. The chart-form constructs developed from this data collection were aligned with those shown by the LLNL Director. Lacy (1992) defines a sequence of analysis list definitions and comparisons. The central reason is given as a plan to develop higher quality and reduced risk. In other words, the reasoning for analysis for the Lacy methodology is to define all of the important characteristics, which in turn allow clear definition for improved quality and reduced risk.

The purpose of using this methodology is to help develop analysis and planning which can improve the quality, and optimize productivity of all of the tasks, and subsequently reduce risk for LLNL.

Critical Project Purpose

It should be intuitively understood that, if the overall LLNL mission and research operations are not clearly defined, then there may be a reduction in quality and a subsequent increase in risk for the end product and in the research process. Lacy's (1992) fundamental list begins with the problem definition, value system design, function analysis, and system synthesis. This allows reduction of risk. Of the seven categories of the research aggregate, the technology assessment provides the greatest detail. This detail led to the discovery of the LLNL ability to integrate many technology areas for advanced research.

This discovery led to the finding and creation of the Concept Fusion model, which is outlined in Chapter 4 and detailed in Appendix A. Given the detail of the technology assessment, the form used is detailed below. The remaining aggregate categories 5, 6, and 7 follow the technology assessment methodology detailed next.

Technology Assessment

A technology assessment was developed for an overall program and project area survey, and innovation and patent development components for the LLNL case study. Each technology field was surveyed for its effect on strategic management. A large group of projects was surveyed which includes samples from most of the research areas found at LLNL. Direct nuclear weapon projects which are classified were not surveyed. However, many of the direct nuclear weapon projects which are not classified provide support for that area, but are also used for socially beneficial project developments. These were surveyed for their interrelationship characteristics. Most of the projects at LLNN are not classified.

For this element of the case study, the technology type and interrelationships were surveyed to provide a base line score or measure, and data comparison of the technology found at LLNL. Technology management principles are found in Chapter 4 for analysis of the multiple interrelationships of the projects which were surveyed.

An assortment of characteristic technology texts, handbooks, and journal articles were surveyed in this component to ascertain the technology placement in the national community. In addition, comparisons to a Euro-community technology assessment were included.

The LLNL is involved m varied forms of device development which relate to products of a diverse nature. On the one hand, one must consider the potential for market value. On the other hand, one must consider the relationship for probable overall investment gain and development potential for the US.

The technology comparisons were in relation to the European Community conference reports as noted in the literature review of Chapter 2, and journal articles cited. Engineering handbooks, as well as technical journal data as in Mechatronics (noted in the literature review), were used for a technical level base line score or measure. The data were structured into the aggregate for consideration of the strategic operations plan and mission. This assessment of the LLNL R&D activities was developed into the aggregate for the overall strategy.

The inclusion of those data, in the evaluation aggregate, is assessed to be the elements for the key assets noted in the chart-forms developed with regard to the Strategic Management Data Composite (item 2) element noted above. Although a critique of each technical area is outside the scope of this project, key technology data were considered. This is noted earlier as a search for the technology area boundaries.

This is the collection of limiting barriers in a technology. This information was used in the combined systems and operations analysis components of item 5 and item 6 of the LLNL analysis aggregate, and contains the issues for item 2.

Survey Analysis Rules

1. Each project is scored according to the face value of the data in the published
   project reports. Whether they may be updated photos, or data from LLNL or any
   source, does not affect the score. If there is other data about the project that are
   not available, then it does not count. In other words, LLNL may have later
   modified the final product, but this assessment uses the data as a benchmark at
   the time of collection. The base line score, is not an evaluation score. This is not
   an attempt to evaluate their projects. It is research and analysis to discover their

2. Each project is compared to data that is found on research and projects not
   accomplished by LLNL. This was not a comparison to other national
   laboratories. However, journal articles, texts, and LLNL cross data are
   compared. Global research activity was surveyed to compare the LLNL
   projects. The scoring was considered from a central starting point of average.
   An average score of 50 is used for the base line score comparisons. The
   average of 50 is with regard to an average dynamic key research agency or
   corporation. It should be understood that meeting this average requires a major
   effort and investment. To provide for plus or minus values to this median, it
   was necessary to find information or issues that could detract from or show its
   superiority of technology. This was accomplished by comparing the project to

Survey rules Continued

known technology levels from journals or text books. This was compared on a global scale. In some instances, this was obvious from the face value. For example, if the project resulted in patenting, this added to the score. These plus and minus item types are shown on the project survey form scoring area.

3. Although there is a scoring, it must be understood that this is subjective. The final results from the overall collection of project evaluations showed compounded and collected values which are not accountable from scormg. Specifically, it is shown in Part 3 and Part 4 that the profound interrelationships of projects to each other provides the basis of advanced innovation production capability.

4. The point system noted in item two is used to help measure a change. This point system is manipulated later in the systems and operations analysis component of Chapter 4, Part 3 for the experimental effect. This is done by evaluation of minimum and maximum values through linear programming model. Data used for points are drawn from an accumulated LLNL report group. In addition, text and journal references are cited for comparison to help provide a base line score. Published LLNL claims about a project are used as scoring device.

Survey Rules Continued

Rule 4. continued

If there is a claim made about a project, but it may not meet the

claim totally, points are subtracted. Although the reason for the difference in

the claim may not be known, it is used as a means to help provide a base line

score. It is expected that, out of a base line score method for 160 projects, a

pattern could emerge which suggests that the overall skills may be in that

range. But the primary issue is a detailing for the discovery that the project

exists sufficiently to be accounted for. This is an emulation of the methods

from accounting referred to as a base line account data. In the final analysis,

it was discovered that the LLNL score seems high compared to a national or

global technology average. More importantly, the survey showed that the

primary issue is of the volume of types and diversity of projects that LLNL is

involved in simultaneously. The score is not the only issue.

All scores are in 5 point increments. No more than 10 points are given for a

data element. This is when there seems to be a duplicate occurrence. But a

maximum of 40 is empirically determined, this is a Base Line Score estimate

for accounting the project, and is not an evaluation. The point item

types are noted below

Survey Rules Continued

Rule 4 continued -the scoring

PLUS:_____MINUS:

Unique in the world Standards issue (per global technology)
Patented Function issue (there may be failure points)
New Safety issue (small—not large issues)
Innovation Claim issue (may not be as good as stated)

For a background understanding, it should be known that this researcher has more

than 7 years direct experience in engineering design evaluations. This includes

evaluations of military and non military products, and associated development
machinery,

in large and small programs, on a national scale. Some of that experience includes some

design evaluation activity at LLNL. In addition, this researcher has technical education

in more than five categories of engineering sciences. This allowed a broad base toward

seeking comparative technology on a global scale, for the complex LLNL engineering

sciences operations. This provided specific skill in the critique of design failure points. It

must be understood that this experience is also sufficient to caution that the scoring must

still be considered subjective. As noted above, the scoring is not an attempt to evaluate

the LLNL projects. It provides an element in the aggregate of data to examine the LLNL

mission.

Explanation of the Forms and Charts in the Assessment

All of the scoring and tabulation data entered into the PROJECT SURVEY FORM was transferred into TABLE 1 for raw data tabulation. The PROJECT SURVEY FORM, although outlined here, is part of the project analysis portfolio. This is the unpublished raw data that is used for TABLE 1 and all subsequent analysis areas. TABLE 1 and the PROJECT SURVEY FORM have the same raw data entry list. This is described below.

Table 1: This table is comprised of a numerically sequential list of the projects surveyed.

This data summary is an accumulation of the data that is found in the survey for each project. Column headings list either abbreviations or acronyms given to simplify the data transfer from the survey form to table 1. The column abbreviations noted below are from left to right on the table. The column items have two significant groups plus an Base Line Score tabulation number.

Table 1: Group one

ID: This is the project number for the table. OPERATION: The engineering sciences general or specific category. PROGRAM: The overall engineering sciences research and development category. LABORATORY: The specific research concentration area. This is usually found to have a specific work and equipment area with a building designation. THRUST AREA: A research category that focuses on a topic area. This is not usually found to require specific laboratory.

148

Evaluation Chart continued

Table 1 Group one continued

However, engineering sciences researchers, along with computer equipment and office structure, define the general work area PROJECTS: The specific project that is used for the survey item. (Name conversions were given)

TABLE 1: Group Two:

The Yes/No Data and Project Survey Base Line Score. DOE: Is this project directed for specific Department of Energy research? In particular, is it for use in energy research, or for ecology, or nuclear waste management ? MIL: Is this project directed for specific Military research and development needs? It is found that many of the projects are not for nuclear weapons research. MIXDIV: Is this project for use in multiple divisions? For example, is the need for Military and non military, or Biomedical and other divisions ? MULTICAT: Does the project need multiple categories of engineering sciences research? For example, project that is found in the technology area of "Optoelectronics" designates a minimum of two categories of overlapping engineering research activity.

PATENT: Is a patent applied for, or does one exist, that was due to inventive work

in the project area ? It is found that LLNL has restrictive rules for this development. Specifically, any patent must grow out of the work on a specific LLNL project. NEW: Can the project area of research be defined as significantly new? In other words, though a patent.

148

Evaluation chart continued

TABLE 1, group two continued

Specifically, any patent must grow out of the work on a specific

LLNL project. NEW: Can the project area of research be defined as

Significantly new? In other words, though a patent may not be obtained, is the

work significantly beyond or out of the areas that are found in the assessment

data search. COOP: Does this project draw from or enhance activities currently

underway at other laboratories? This is not a private enterprise project.

CRADA: Is there a Cooperative Research And Development Agreement

(CRADA) contract noted for this project? The CRADA are contracts with private

enterprise companies or universities. This can be for project that notes a

CRADA, even though the CRADA is no longer in effect. BASE LINE SCORE:

This is the project existence score. This is a first encounter rough score, to

provide a base line utility value. The method is taken from large scale cost

accounting, where it is preferable to at least list the items' (base line) for its

potential cost. In this case, the method is used to survey its potential benefit.

Given that this program is not intended to be an evaluation of the projects, the

base line score provides rough information for the data and numbers collected.

Evaluation cart continued

TABLE 1. group two continued

The large group of projects provides bulk information for studying the

In comparing the statistical scores, to the subcategorical information (YES/NO)

data, it was found that the YES/NO information provides the predominant

information to study the mission. Therefore, no further iterations for the base

line score were needed.REQ: Is this project mandated by specific national

security needs? In other words, the project is most likely part of the nuclear

weapons stockpile maintenance needs. This is for general information.

However, the lack of a yes does not mean the project is not required. For

example, an "Optoelectronic" project requires both Electronics and Photonics

research areas. These forms of projects are found to evoke a mixed division

definition.

The Project Survey Form

This form is designed for the survey of each of the projects. One 8 1/2x11 sheet

comprises the primary survey form. It is a sample, but is also the survey for project

number 144. This is a survey sensor project, which was given the acronym name of

OPTOELECT. This particular sensor is comprised of both optical and electronic systems.

It relates to projects that are found in the Society of Photooptical and Instrumentation

Engineers (SPIE) journals.

Therefore, the score would be average in high technology parameters.

This, of course, must be understood as a survey which seeks quick comparative

information for an accounting of the project area. Additional sheets are added where the

explanation detail involves complicated technological description. There is one Project

Survey Form for each project. This Technology Assessment includes over 160 of these

forms for the study of the LLNL mission. These are retained in the project portfolio, but

are not required for the final presentation of this project. Each of the items noted in this

form is transferred into Table 1, noted above. The significant difference in this form, to

the list of tabular columns, is found in the base line scoring and tabulation blocks. These

are found just below the project items.

Base Line Score Tabulation Blocks

There are three significant blocks provided which define the scoring data

collection. These are: the Plus score group; the Minus score group; and the overall score

tabulation box. Just below the total score tabulation box, the survey base line total score

is shown. This base line score is transferred to the base line score column in Table 1. The

yes/no data for each project is also transferred into the appropriate columns in Table 1.

PROJECT SURVEY FORM

PROJECT REPORT NUMBER ID: PROJECTS: ESTEST
DIVISION: |XYZ   shown below.

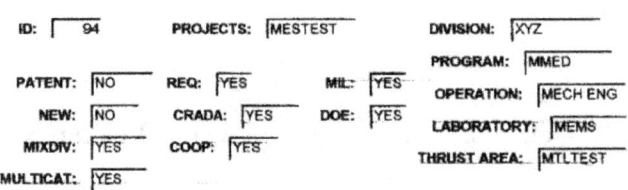

DIVISION ACTIVITY DATA AREA

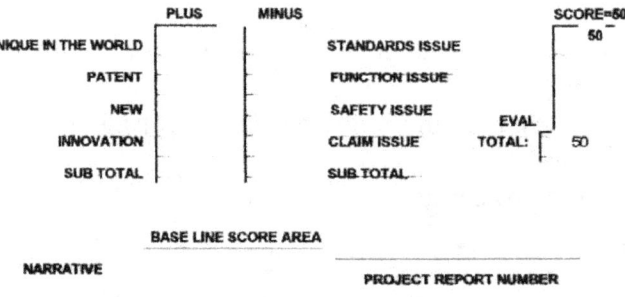

BASE LINE SCORE AREA

NARRATIVE

PROJECT REPORT NUMBER

Ultra high precision measurement technology and systems are noted at LLNL. Eight significant device systems are noted. The combination provides a basis for mixed division needs. This project can be viewed as a work volume area.

This project survey is of a materials stress test station and system. It features advanced instrumentation systems. This is seen as an average functional test system for research purposes. There were no obvious questions about the project, nor any extremely unique characteristics. Therefore, the average of 50 points is suggested. However, the adaptation of the technology exhibited the value to be used in a mix of divisions, for both military and non military requirements. Multiple categories of engineering sciences are required for this test system, because mechanical engineering is interfaced to electronic instrumentation and measurement systems.

Figure 1. Project Survey Form

FIGURE ONE   and   NARRATIVE

Example of all forms used for the survey.

Information Sciences Strategic Plan

A survey of the overall planning for the LLNL information sciences was shown to be part of the LLNL aggregate. Chapter 2 reference (CD-ROMs in libraries) provides information for a base line measure. In addition, projects discovered in the Technology Assessment component demonstrated the LLNL is advanced and has advanced planning in this area. For consideration of future developments, Toffler (1990, p. 259) refers to the information wars of today. Toffler's concept and projected trend shows Info-Tactics as essential for staying current and accurate in this changing world social system. To follow this approach, a topical survey uses data from the technology survey of the other project areas. In other words, innovation development efficiency shows data for information change velocity. In addition, the aggregate of overall strategic management planning defines the velocity and characteristics for information needs for that strategy.

In summary, the information sciences aspect of the overall analysis is a critical element for every aspect of the LLNL future project efficiency and viability. The overall case study was developed to consider this cross comparative needs for the Chartforms and LLNL program plans show a critical need and central development strategy, and also show primary elements for an information quality assessment. The primary issue is that the overall capability of LLNL is not well understood. The capability, and therefore the quality, was summarily shown to be high in this assessment, because the information systems were shown to be an integrated component of the overall operations planning.

This aggregate component, therefore, became a structured component of the systems and operations analysis composite of items 2, 3, 5, and 6.

Operations Analysis of Engineering Sciences

Operations analysis of the specific programs and projects was developed as part of the aggregate evaluation. Some of the prior developed systemic chart-form data, noted in item 2, showed an overlap and direct connectivity of the elements of this aggregate analysis component. Therefore, this component became an imbedded systems and operations analysis section composite of Chapter 4. The Strategic management composite of item 2 also became embedded in this component because of the effective overlap of planning demonstrated by LLNL. For example: given the overall program plans defined for the LLNL yearly activities, given the personnel and departmental organizational plan, given the project definition, given the inclusion of simultaneous CRADA operations and other data, the overall projected innovation performance was surveyed. The data showed that the primary issues for understanding the mission rests with the understanding of knowledge production and subsequent innovation production. The issues of employee shifting are understood more fully when analyzing the issues and methods for innovation production. LLNL demonstrates a composite operations plan. This is examined in the systems and operations analysis component composite of Chapter 4.

Technology Region Survey

Regional planning for the San Francisco bay-area was searched for this project. This was envisioned as comparable to the planning, defined as research centers, noted by the references found on the European Community (EC). This was considered for the overall changes found in the bay-area.

The issue of agglomeration, pinpointed by Hanson (1996) was considered for

LLNL as placed in the San Francisco Bay Area. The potential for analysis questions is of

a collection of national laboratories,

Silicone Valley, and the pervading terrain of the Bay Area. Regional planning data

for recent and current California and local planning was considered, but was discovered

not to be a significant factor for this case study. The LLNL has developed research data

for their placement in the region. It was discovered that the primary issue for this project

centers on the issues of knowledge growth and innovation productivity. This is not a

regional issue in this case.

Summary

The overall research methodology comprised a set of data collection devices which provided triangular validation of the data for each. This method is used by the General Accounting Office (GAO) for special case studies. The limitation of the GAO research, shown in current government reports, is that the GAO approach does not involve analysis of the fundamental principles of operation that a national laboratory needs for knowledge and innovation production.

Each component of this research project provides a collection of data and comparison material. Although a few of the elements in the data collection did not yield the pivotal material as others did, the overall case study considerations of the material proved to be enlightening toward the primary issues about the LLNL mission. The primary issues were discovered. These are best understood as a composite. This is because, complete answers on complex research topics, are not answerable by noncomplex analysis. The complex composite provides cross validation of the needs for the LLNL mission. In the final analysis, it is shown that immense innovation capability exists using the management and aggregate research methods found at LLNL.

The primary project was to discover the definition of the actual mission of LLNL. This is suggested to represent a paradox. It represents the development potential for many advanced sciences that are beneficial to society, while continuing analysis of the nuclear weapons stockpile for safe storage.

Systems and Operations Questions
Regarding the Engineering Sciences Operations Policy and Planning

A list of general questions about the LLNL operations and mission follows. It is designed to focus on significant issues which are answered from this research project. The answers to the questions are found in Chapter 5. The general questions are to help determine a placement of LLNL in the national community of activities.

1. Do the operations follow norms found in non government entities—industry? Can it be said that the operations parallel business ?

2. Do the operations experience catastrophic developments due to changes in Government policy? Or, does the operations design allow proportionate changes? Can a change impact be compared to market changes for industry? Is LLNL able to respond to changes effectively?

3. Does the military project research planning allow technology adaptation for nonmilitary use? Does the operation plan lend toward a clear duel use benefit?

4. Does the State of California provide financial support of the LLNL research operations?

5. Are systemic needs internally consistent—toward efficiency and creativity?

6. Do information sciences technologies and planning permeate the operations?

7. Given a known nuclear weapons responsibility, can one find a central function which is not actually nuclear weapons oriented?

8. Do technology skill areas seem to operate at the same quality level in all activity areas?

9. Does the LLNL location appear to be appropriate? Can significant location questions be found?

10. Do the operations provide benefit to national education?

11. Can technology research areas be found which may be effectively divested or added?

Answers to the above questions are expected to be general. Some questions may produce a set of added questions which could be expected to provide useful information for futures planning and research.

CHAPTER 4

RESULTS
Overview of the Research Analysis

The following analysis for this case study research of the LLNL mission was performed. An aggregate of research categories were created toward that purpose. Seven categories of analysis were conducted: General Operations Analysis, Strategic Management Data Composite, Systemic Structural Analysis, echnology Assessment. Information Sciences Strategic Plan.

Operations Analysis of Engineering Sciences.

Technology Region Survey.

The research determined that the methods practiced in the Engineering Sciences Operation include overlapping and integrated management analysis. Therefore, the seven items noted above are now included in three parts. Part 1 is the Technology Region Survey (7). Part 2 is the Technology Assessment (4). This follows the Technology region survey. Part 3 is the Systems and Operations Analysis, which includes an aggregation of four of the aggregate components listed in Chapter 3, and noted above. These are: General Operations Analysis (1), Systemic Structural Analysis (3), Operations Analysis of Engineering Sciences (6), and the Information Sciences Strategic Plan (5).

Part 4 examines the fundamental principles of operations that occur in an ideal research agency for high productivity of knowledge and innovation production on a national scale. This research and analysis shows that the Lawrence Livermore National Laboratory (LLNL) meets the requirements for an ideal research agency, for the benefit of the United States.

The fourth part is defined as a Concept Fusion model (Cutburth, 1997). An outline of the contents of Parts 1, 2, 3 and 4 follows.

This research, its method, and scoring were reviewed and validated by the LLNL directorate. It is stipulated that the project assessment scoring method provides general information, but is changeable due to the ongoing changes in the projects and information available. The general scoring is collected using a base line score method, described in Part 2. This provides a flexible determination of the score value, depending on the overall LLNL score and information.

The essential LLNL validation is about the existence, interactivity, and capabilities that this research project discovered. The Technology Region Survey comprises a brief summary of literature surveyed about the location of LLNL. There were no significant issues discovered which would impact the mission of LLNL.

The Technology Assessment of the Engineering Sciences Operations is the collection and survey of over 160 project areas that are being or have been developed in the engineering sciences operations at LLNL.

Project survey forms were constructed which provide key information on each project. The primary value of the information is gathered to provide significant overview and confirmation that a project exists, the form or type of project, and its connection to the other projects, that are typical of the requirements of the engineering sciences operations.

A base line score is used. The base line score is not an evaluation score. This is not an attempt to evaluate their projects, but an attempt to discover LLNL's mission. The scoring is to determine the existence of the project, and its possible placement in the operations. However, the scoring is subjective, and must be considered informational only. The base line score, or measure, is to be viewed as a form of preliminary audit whereby it is preferred to have some data about the subject as opposed to no accounting for it.

The Base Line Score is subject to change if new or more detailed data offsets the number. In the summary, after analysis, it was found that no new detailed information was required for the Base Line Scores, because the key issue is the existence of the broad and intense diversity, and connectivity of forms of research. Specifically, the accounting of the project types and diversity provided the key data for understanding the LLNL mission.

162

The Systems and Operations Analysis of the Engineering Sciences Operations, includes a collection of several of the analysis synthesis components that are listed in Chapter 3—the methodology. These are: General Operations Analysis, Systemic Structural Analysis, Operations Analysis of Engineering Sciences, and the Information Sciences Strategic Plan. A complex aggregate of text and book ideas are examined for analysis of the mission. A set of goal choices for the engineering sciences operations were

experimentally examined using Saaty (1995), Expert Choice) software. In addition, the overall operations personnel shifting was experimentally examined using linear programming software and text methods by

Schrage (1996). The software used has the acronym LINDO, and is used for linear programming analysis. In addition, Dynamic Programming software by the Lantana corporation, Vensim Molecules, was used to construct a model for consideration of a closed loop process. This closed loop analysis is used to develop an operations plan for the engineering sciences operations.

It is demonstrated that, although the software and analysis methods offer development and linear manipulation of linear variables, the operations at LLNL demonstrate nonlinear variable systems. In particular, the generation of knowledge and prediction of innovation are shown to present nonlinear, and stochastic differential systems.

The Concept Fusion model (Cutburth, 1997) demonstrates that the LLNL operations provide a basis for a knowledge and innovation production mission. The model demonstrates that standard management analysis methods are not appropriate, but can be expanded to optimize productivity and reduce knowledge and funding waste. The significant discoveries and analysis are drawn from evaluation of a theory of knowledge growth and subsequent innovation production. The technical references in the literature review and other references with regard to theories of knowledge, mathematics, neuroscience, and plasma physics are aggregated in the analysis of theoretical knowledge production concepts.

Theoretical mathematical principles that are directly connected to language are surveyed to demonstrate the existence of a pure mathematical linguistic, and an a priori (preexistent) mental computation skill of the human brain. This is used to demonstrate a knowledge growth potential that emulates a nuclear fusion process. The knowledge growth potential is both for the individuals in and the social organization of, an ideal research agency. The existence of an innovation science is suggested. Part 4 provides a basis of future research needs.

PART 1. Technology Region Survey

The Lawrence Livermore National Laboratory (LLNL) is located near Livermore, California. The overall laboratory has two primary locations. These are both situated around and in a mountainous area between Livermore and Tracy, California. The setting is centered about 70 miles east of San Francisco. Two other national laboratories have facilities near LLNL. Lawrence Berkeley National Laboratory (LBL) is located in Berkeley, California, adjacent to the University of California in Berkeley, around 45 miles from LLNL. The Sandia National Laboratory (SNL) has a pilot facility located across the road from LLNL. The SNL primary facilities are located in New Mexico, along with the primary facility of Los Alamos National Laboratory (LANL).

The world-renowned Silicon Valley is situated approximately 50 miles from LLNL. Stanford University, which conducts nuclear physics research, is located approximately 70 miles from LLNL. In other words, LLNL is located in an exceptionally high technology region. The LLNL has developed significant data with regard to its functional region, but a cross comparison to published literature is considered below.

Although comparisons may be helpful for an overview, one should examine the literature offered by LLNL. It was discovered through analysis of LLNL reports that LLNL has an internal engineering sciences operations plan that is substantially self-dependent. It is shown further that the operations and management planning at LLNL can generate high innovation output due to this plan.

This provides benefit to any research entity which may operate in cooperation with LLNL. The focus on this project is analysis of an advanced research agency. The selfconsistency found in the LLNL operations plan indicates that a high innovation yield, even if the location region were less ideal. In other words, although one can examine the ideal value for the extremely advanced technology region which LLNL is positioned within, the analysis components below indicate that the value of LLNL is shown through its internal operational consistency and efficiency. Specifically, LLNL is shown to be an ideal research agency by means of its internal operations. Therefore, the region of location did not become the focus needed to examine the most productive innovation plan for an ideal research agency.

The reference texts and journal articles cited in the literature review of chapter 2 provide examples for analysis of regional considerations. Of most particular interest are those with regard to international competition. Science parks and technology region analysis are comparative for the external effects on innovation productivity on a large scale.

Summerton (1994) examined international product changes for the European community. A key idea she suggested is that a product should not be viewed as an isolated artifact. An example was given that home telephone is not a simple home appliance'; it represents an interconnected social network and technical products.

LLNL demonstrates that a singular object may have multiple categories of technology needs, and a self-consistent agency may develop the entire sphere of the technologies. In addition, these may evolve into high level products within one advanced research agency. Therefore, although Summerton (1994) pointed to a significant concept, this does not prevent internal connected development at one agency. LLNL demonstrates the unique ability to achieve high levels of advancement toward interconnected research that is beneficial to efficiency. LLNL research and development of sensor products provide examples. The development facts of these are delineated in the Technology Assessment, Part 2.

The interconnectivity of the LLNL sensor product development is considered in Part 2, 3, and the Concept Fusion model (Cutburth, 1997). Henry (1991) examined the abilities to forecast technology innovation, and did so by examining the European community. However, he concluded that the prediction of innovation is statistically nearly zero. When examining the overall LLNL funding, in each category, funding was shown to produce innovation. Although a re-examination of the conclusion by Henry is outside the scope of this project,

Henry's statistic was based on the combined European Community funding results, and strictly contrastive to the results shown at LLNL. A strict difference is shown when examining Parts 3 and 4 below. It is shown that an aggregate program at one agency allows high innovation potential.

Fusfeld and Haklisch (1984) examined the industry-university interaction for research. They argue that the interaction is beneficial toward simultaneous industrial advancement and education. The LLNL is actually governed by a body at the University of California. Furthermore, LLNL was founded in this fashion. However, LLNL has demonstrated a self-consistent development plan which simultaneously produces knowledge advancement and innovation productivity that can allow product development. In other words, though LLNL is associated by legal foundation to the University of California, the Technology Assessment demonstrated simultaneous product productivity and knowledge growth. This is shown to be accelerated by the combined LLNL management plan, and the revelation from the Concept Fusion model demonstrated in this project.

Fusfeld and Haklisch (1984) argued that, unless there is university and industrial cooperation, education will fall behind industry. However, in examining the self-consistent programs for knowledge growth and "Information Engineering" practiced at LLNL, one argues a rationale that can simultaneously accelerate industrial and educational developments. Notwithstanding that LLNL was founded in a university contract with the government, LLNL can be said to have evolved into a selfconsistent research agency—that is offset from the Fusfeld and Haklisch (1984) plan.Therefore, although outside the scope of this project, one can consider whether the Fusfeld and Haklisch (1984) theory could sponsor an entity that may evolve into something that emulates the LLNL plan.

Subsequently, it is argued that the focus on understanding the actual needs of an ideal research agency, as done in this project, may contribute more rapidly to a conclusion on the most universal plan.

Monck (1988) examined a combination that is similar to the Fusfeld and Haklisch (1984) plan, but added the government component. Of course, LLNL is funded largely by government agencies, but it is argued that the evolution would eventually produce the final results shown in this study. Cross comparison of possible combinations considered by both Monck (1988) and Fusfeld and Haklisch (1984) provides a form of triangular analysis of concepts. But it is shown in the components in this project, that a triangular validation process is precisely that which produces knowledge—and this is what is practiced at LLNL.

Therefore, it is argued that the most ideal research agency is that which evolves a plan that emulates that found at LLNL. Specifically, Monck (1988) and Fusfeld and Haklisch (1984) analysis comparisons are attempts to examine—and to search for—the ideal research agency toward the highest innovation yield. However, a definition of knowledge was not well defined in the studies, giving question about the fundamental measure.

In conclusion, a Technology Region Survey for this project helped validate the primary research focus or this project. The focus is found in discovery of the needs for an ideal research agency, which should be transparent to the planning which sponsors its creation. Furthermore, although the region in which LLNL resides is extremely advanced in technology, LLNL demonstrates operations efficiency that appears somewhat self-sufficient. This is shown in Parts 2 through 4 to be ideal for advanced innovation production. The research from this aggregate component revealed that the internal functions of LLNL offer somewhat stable analysis, because LLNL is somewhat self-sufficient. The examination of LLNL projects surveyed in Part 2 provide a perspective which shows consistency in research and development planning.

This is also shown in Parts 3 and 4, to provide a basis of profound growth potential in innovation production.

PART 2. Technology Assessment of the Engineering Science Operations

Technology assessment is one of seven categories of analysis of the engineering sciences operations. The seven categories provide an aggregate of data for a case study. The aggregate concept is defined by the General Accounting Office (GAO) in their text on an Evaluation Synthesis GAO/PEMD-10.1.2.

This GAO regulation teaches that construction of an aggregate of data provides a means for comprehensive evaluation. For this project, its assessment is denned as an overall summary of data and item survey values. The data are from surveys of projects in the technology categories of research found at Lawrence Livermore National Laboratory (LLNL). The surveys were made of published LLNL reports. There are both benefits and limitations to the overall assessment which are explained below. In addition, the basic rules for (analysis) and assessment are defined. It is also shown that this Technology

Assessment is comprised of an aggregate of data and item surveys. In other words, an aggregate of technology areas is found to exist at LLNL. Furthermore, an aggregate analysis methodology is shown to be ideal for analysis of the LLNL engineering sciences operations, for discovery of its mission.

The analysis aggregate includes a survey of over 160 projects found at LLNL. The selection is of projects which describe most of the research topic areas that LLNL is specifically involved in. These are listed in TABLE 1. In addition, a general Base Line Score (see Part 1 of this chapter) is given to each of the projects. Supportive data are assembled for each of the projects, along with a survey form and project cover page. The data are valuable for use in the systems and operations analysis component, for a theoretical determination of the mission. Therefore, the data are used, but the cover pages are omitted in this project.

The supportive data for each project are comprised of LLNL published reports. All published reports used for the survey are available from LLNL, but not included in this dissertation. It is shown that the overall development of an aggregate, provides a unique understanding of the LLNL engineering sciences operations. Furthermore, the experimental analysis, using the overall aggregate data, provides significant material for an improved understanding of the LLNL mission.

Benefits and Limitations

1. LLNL is involved in projects, some of which are legally defined as CLASSIFIED. This creates some limitation as to what projects can be listed for those surveyed. Therefore, a large volume of nonclassified projects are survey, to provide detail for

   this research.

1. Although some projects are classified, it is found that the bulk of the LLNL engineering sciences operations work is not classified. Therefore, a bulk analysis, using an aggregate methodology, provides useful data for this overall LLNL Case

PART 3   Systems and Operations Analysis
of the Engineering Sciences Operations

This component uses an aggregate of information for analysis of the Lawrence Livermore National Laboratory (LLNL) Engineering Sciences Operations mission. This aggregate is part of the overall project analysis aggregate for analysis of the LLNL mission. The primary analysis is of the Engineering Sciences Operations (ESO) and its mission, which is assigned a primary task of Research and

Development to actualize the LLNL mission. Data for this analysis are from the following sources:

1. The Technology Assessment (TA) component of this project.

2. Information acquisition which developed from interviews of the Engineering Sciences Operation Director (ESOD).

3. LLNL Comptroller and Engineering Sciences Operations Director yearly reports. See UCRL-AR- 125806, and UCRL-ID-127001.

To develop an aggregate of information about the LLNL mission, a selection of various analysis methods were experimentally used here. Each distinct method was used to provide information from variety of perspectives.

This allowed a triangular validation of the information discovered, by use of the others. The information is added to the information discovered in all of the other analysis aggregate methods noted in the methodology of chapter 3. The complete aggregate is used, for the conclusions from the research, that is detailed in Chapter 5.

Decision making computer software, for experimental analysis of mission components and choices. The Expert Choice decision software is used to analyze the LLNL strategic goal choice set, for analysis of its mission. The model automatically computes comparative statistical values from variations in data that are input from choices that are made on questionnaire forms.

Linear Programming computer software was used for experimental analysis of program variations. LINDO Operations analysis (linear programming) software was used to examine the effects of shifting large groups of employees from one research program to another. This application of the LINDO model was used for finding equilibrium when large changes occur from large changes in national funding. This is adapted to historical changes that have occurred, to examine what input and output would be useful for analysis of the LLNL research objectives, to develop the LLNL

Systems analysis text model comparisons with the Engineering Sciences Operations plans. The systems analysis defined by Lacy (1992) was used to define research information pertaining to issues or situations that develop in an organization. This technology, supported by text methods was used to seek some form of systemicy in the research planning, toward development of operations models.

Systemic analysis and cyclical dynamic operations model development, using operations model computer software. Vensim Molecules operations research model software was used for this. This software allows dynamic variable comparisons and/or computation from input of many multiple variables that are found in an operational management system. This is for developing experimental models of the Engineering

Sciences Operations mission and LLNL mission model. These are shown in graphic forms. One is titled Concept Fusion model, and one is titled Engineering Sciences Operations Plan.Decision sciences text model comparisons are made with the Engineering Sciences Operations plans. A group of management analysis text methods are compared for the information they provide. They are compared to the information found m the LLNL reports on their operations and

The experimental application of the above models provided an understanding of the complicated operations requirements of LLNL. Many unique evaluation techniques exist for operations analysis. Some have been noted in the literature review of chapter 2. They include methods used in analysis of large scale and multinational corporations. In addition to the texts, new software analysis methods, as noted above, were experimentally used to broaden the understanding for their possible application in high technology research activities.

These demonstrated that the LLNL operations planning has unique combined methods which can be used for any large scale research program. Except that the analysis requirements of LLNL are more complex, exhibiting stochastic differential systems that standard business analysis methods are generally not designed to analyze. The methods do not specifically detail analysis of knowledge or innovation production. This component begins with a report on the information obtained in the interview and conversations with the ESOD. It was discovered that a flexible plan of operations is at the basis of all operations planning at LLNL. The flexibility is shown to provide ideas on how to run a high technology research organization. In addition to the overall flexibility planning found at LLNL, a historical concept for flexibility seems to be applied for recent changes to the strategic planning of the ESOD.

Specifically, a Thrust Area realignment was undertaken that is along the lines of flexible planning methodology. An interpretation of the change is discussed below. Although this demonstrates a characteristic change which centers on the original LLNL flexible planning methodology, the significant point is that LLNL is transforming in its strategic structure to align with transformations in technology. Other transformations can be expected.

A unique broad and interactive research program was discovered in the Technology Assessment (TA) component of this study. The interactivity of the programs found in the TA can be described as intense, and continuous. The existence of the interactivity allows an expansion on a new concept that is explored below, titled Concept Fusion (Cutburth, 1997).

An aggregate approach is used for discovery of information in this analysis. A unique comparison was found by this method. This is that LLNL practices application of an aggregate of research programs for optimization of innovation productivity. Flexible planning methods were also found which augment the aggregation and productivity.

It is suggested that the LLNL research aggregate is functionally ideal, and explained partly in the description of this CF model. It is suggested that an innovation science can exist through adaptation and practice of concepts in this model.

Interview Report

This report is of an interview with the Associate Director of LLNL, that is in charge of the Engineering Sciences Operations. This must not be seen as an interview of the person, but of an LLNL co-director, who was able to define the essential operations plans. The interview of the associate director and his technical assistant totaled approximately 3 hours. In addition, there was a half hour follow up meeting with the associate director to outline the data discovered in the technology assessment portion of this research. The guidance given by the associate director was strategic. However, this is not an analysis or evaluation of the persons or employees at LLNL.

This research project is for the analysis of the Engineering Sciences Operations, for analysis of the LLNL The Engineering Sciences Operations (ESOD) division includes more than 2,100 full time employees (FTE). The division provides engineering support for all of the research activities at LLNL. An organization chart is shown in Figure 2. This is replicated from the ESOD report for 1996, provided by LLNL, through the ESOD. (See the full Engineering Annual Summary UCRL-ID-127001.) The interview was accompanied by phone conversations. The meeting was attended by this researcher, the director, and one director assistant/division specialist, and took place at LLNL in November 1996. In addition to the meeting, the director delegated responsibility to area leaders to provide significant reference information for this project, in the form of published LLNL reports.

A follow-up meeting for review of the Technology Assessment (TA) component was held in April 1997 at LLNL. The TA provides general information about the overall program, division, and thrust area R&D activities. It should be understood that the TA provides general, but not total information about the overall R&D conducted at LLNL. This research objective, to seek engineering sciences operations concepts for an advanced research agency, has been fulfilled by those meetings and subsequent published report information transfer, and this analysis.

The original interview meeting provided significant strategic operations instruction for this researcher. Conversations and hand sketches from the ESOD provided an overview of the strategic planning by LLNL and the ESOD. In addition, the specialist, who has many years of LLNL research management experience, was able to provide relationships to past planning. The meeting revealed an effective methodology in LLNL planning.

Figure 2. Operations 1996   EngineeringSciencesChart131 Organization

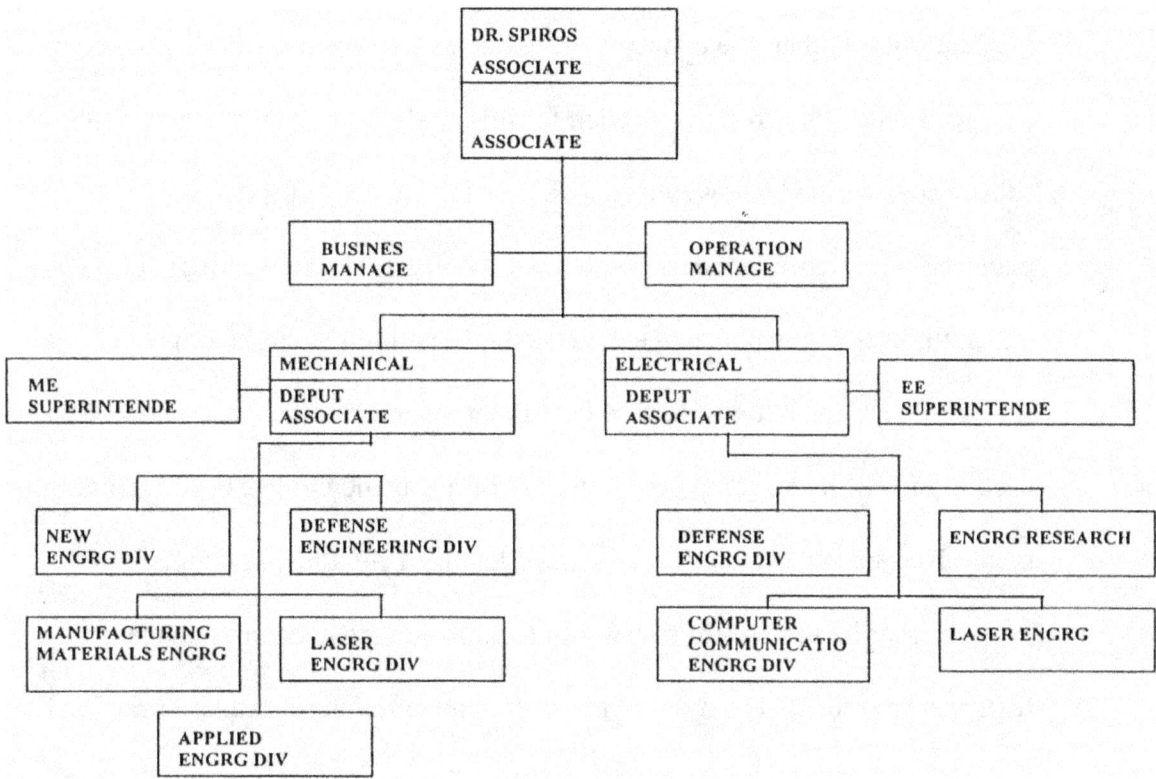

There is no attempt to define a direct quote of comments by the director or the specialist. Rather, the concepts presented and subsequent published report survey provide the substance needed for this study. The interpretation of the substance is defined below, along with general sketches of the program planning. The significant items which were revealed at the meetings follow:

1. LLNL uses a matrix management method in all of the engineering sciences operations. It was further noted that this method was claimed to be used at LLNL since perhaps the founding of the agency in the 1940s. This method has been used to allow advanced levels of flexible planning.

To ascribe some form of validity for this method, current management texts were explored. The method is current and is found in standard practice for corporations involved in multiple programs. For example, Gaynor (1996) provides an analysis using matrix management methods.

It is shown that the LLNL management methods, including the matrix management plan, facilitates advanced mathematical simulation concepts. It is found that most of the advanced mathematical methods used in the sciences and management analysis use matrix computation. The mathematical correspondence is examined below, under the heading of the Concept Fusion (CF) model (Cutburth, 1997).

2. Multiple programs, divisions, and thrust areas are shown in the analysis below to have interactive planning operations. These approaches to R&D were defined as continuously flexible—in an environment of the matrix management method. In other words, all of the categories are assumed flexible and changeable for optimal adaptation toward change. The change potential is that which is deemed automatic due to new technology that the LLNL develops.

   A hand sketch was drawn by the ESOD to give a general pictorial representation of the concepts for change that is possible by the LLNL i interactive matrix method. A duplication is sketched below (see Figure 4). The sketch is similar to drawings provided by Gaynor (1996) on management of technology.

3. Flexible program, division, and thrust area changes were shown to be used

   for analysis of possible reconfiguration in the final ESOD strategic plan for 1996. Current planning was revealed that resulted in redefining, and renaming, of a significant thrust area. The new thrust area is titled Information Engineering (IE). This is planned as an increased focus on improving all communications methods, devices, principles and concepts toward accelerated information transfer—for improved R&D production.

The significant point about this change, is that restructuring or transformation occurs to provide a strategic re-alignment to meet technology transformations. Other changes can be expected. The fundamental management methods, including the matrix and strategic alignment planning system, were shown in relation to development of new significant research areas, as noted below.

4. In addition, the management methods were shown for their adaptation to divestiture of programs, which can be turned over for general public use. For example, The United States Enrichment Corporation (USEC) purchased the concepts and rights to manufacture nuclear reactor fuel. The process is referred to as Advanced Laser Isotope Separation (AVLIS). The process was developed by LLNL. In addition to this sale of a process, LLNL has a standing contract to provide consultation services to the USEC. Although this sale of the multimillion dollar process has been highly publicized, the reasoning with regard to the operations planning was defined in the discussion. The simple sketch (Figure 3a, 3b, and 3c, p. 132) provided by the ESOD are reported in the LLNL publication UCRL-AR-122705-96.

5. There were at least seven forms of product productivity described in the overall management methods discussed, and shown in the reports. These are interpreted by this author and shown below as Type A, B, C, D, E, F, and G.

LLNL R&D of isotope production process, a large scale Type A: AVLIS division. AVLIS R&D began as a program. After exploration and production capability was demonstrated (by development of a complete pilot production facility), the process rights were sold. See LLNL report UCRL-AR-122705-96.

Type B: Development of a significant research focus area (as in Micro-pulsed radar) that grew out of a thrust area research project. The program has been defined recently and accelerated for R&D into new product developments. The Micro-pulsed radar technology and spin-off product group has been growing. This technology growth area can be referred to as from the primary program, in the area of radar R&D. See LLNL report UCRL-52000-96-1/2.

Type C: A third form of productivity was demonstrated as an overall LLNL core skill area. This is with regard to sensor systems R&D, and other optoelectronic devices. Sensor system development was noted as a necessary R&D component to meet the needs of all of the research programs. In addition, it was noted as the development of a core high skill area. The sensor systems were shown to be part of LLNL continuous development (a process that permeates LLNL R&D). In addition, singular products which use sensors were found to provide compounded product development areas. These are subsequently marketed on an item by item basis. See the LLNL report UCRL-TB-119860.

Type D. LLNL Core knowledge growth. This is interpreted as a product by LLNL, because it is the basis which is used by the United States Congress to continuously fill the needs of the US.

Type E. Technical publications about advancements in technology in least 40 diverse areas.

Type F. Consultation services for health sciences, general but advanced scientific products, and international defense systems.

Type G. Public science education.

Sensor systems technology advancement emanates from the large mass of overall programs that exist at LLNL. This is analogous to fusion energy production that is described in the Concept Fusion model, because of the high yield of multiple product capability, and a form of chain reaction development.

Product marketing methods were reported by the ESOD to have two distinct avenues. One is by sale or licensing of spin-off technology. The other is a direct LLNL solicitation of partners for advanced development of discovered technology. This is in the form of contracts—known as Cooperative Research and Development Activity (CRADA).

The ESOD noted that there has been a decline in the CRADA business activity due to reductions in Congressional interest, and constraints for product development defined by standing laws. The significant issue is that LLNL has a longstanding method for this form of marketing, as authorized by Congress.

Except that, as observed by this author, outside funding may not be adequate to optimize the marketing potential of products that LLNL produces. The existence of greater opportunity is demonstrated in the Concept Fusion model (Cutburth, 1997).

Information is found about possible CRADA contracts in the LLNL publication on technology transfer, Opportunities for Partnership, 2nd ed. UCRL-TB-110794-95. However, the overall planning for most of the forms of developments appear to be driven by the requirements of the significant customers of LLNL. These are essentially the Department of Defense (DOD) and the Department of Energy (DOE).

The new requirements of Congress through the DOD are that all research be considered for its multiple application potential. Also, the DOD has directed that military-research be designed to provide direct nonmilitary benefit wherever possible.

The concept of multiple use of research achievements appears to have existed for decades at LLNL. A primary example is in the development of sensor systems and associated computer software. They are applicable for sensing and computing as if they are generic products. The expanded DOD requirement should accelerate the diversity of technology development. On the other hand, compounding technology development can be hampered by funding constraints.

Although the second use could be adapted at reduced cost, it should be understood that the duel use describes possible added cost, in order to optimize both applications.

Figure 3 (p. 137), Figure 4 (p. 138), and Figure 5 (p. 139), are representations of the ESOD sketches of the change process. They are brief. However, they represent an overview of the LLNL ESOD change processes, and should be considered as representing dynamic change. They represent interactive processes due to the complexity of program overlaps. Figure 3.a, 3.b, and 3.c show the overlapping strategic plan. These are interpreted as three relative perspectives of the combined plan. Figure 4 represents a strategic program and project plan. The matrix shows directed arrows for a dual potential for LLNL. The program shifting can become a self sustaining technology, or follow the opposite path toward outsourcing or divestiture.

Figure 3 (3.a, 3.b, 3.c) provides a strategic overview and comparison of LLNL operations to business sizes. The thrust area representation is considered as similar to small business venture activity. This can be sponsored from core technology areas or provide technology growth for core technology areas.

Figures 3a, 3b, and 3c provide three perspectives of the aggregate of technology development at LLNL. Each provides input of items, which provides an output purpose. In 3a, the input is a collection of the institution, people, and administration. The output is the development of the people who are part of the overall plan. In Figure 3.b, the input of current science and technology, programmatic orientation, and planning and program management provides the basis for output when funding is provided.

This collection of the input group is geared to produce large scale technological developments which are comprised of many technologies. Figure 3.c shows a collection of the organization as a hole, which includes the associated multiple disciplines of the individuals in the collection, and the people. The output is the development of highly advanced singular technologies which are usable in many ways. Such as the advanced development of materials which are in turn usable in many product.

Figures 4 and 5 are graphical demonstrations of the strategic plans and their results. They show flexible input and flexible output ideas in the strategic planning.

After review of the interview and materials supplied, a new perspective was conceived for the results of the LLNL and ESOD strategic planning. This is described below, as developed by this author, in two new sketches (Figure 6, and Figure 11). The first is an interpretation of the ESOD strategic plan development process. The second is a theoretical construct to describe a compounding resultant of the original R&D process planning by the LLNL ESOD. A Concept Fusion CF) process (Cutburth, 1997).

Information and fact finding methods for this most recent study began by use of a combination of methods. The GAO case study methods provided the fundamental building block, which is the discovery and use of an aggregate of information to form a triangulation for information validation The GAO method was further augmented by use of a systems analysis information acquisition method.

The systems analysis method by Lacy (1992) is used to seek a system which can be described by development of a systemic model. The models, Figure 6 and Figure 11, were drawn by use of the Vensim-Molecules 1995, Lantana Corp, operations research model building software. The software was loaned to this student for this project by the Lantana Corporation (Hines, Eberlein, Richardson, Johnson, Richmond, Milhish, Ho, 1996).

Hines, Eberlein, Richardson, Johnson, Richmond, Melhish, and Ho (1996) provide the operations research software that is used in the model shown next. This model is referred to as a Strategic Operations Plan (Figure 6, p. 145) for the LLNL engineering sciences operations, for actualization of the LLNL mission. The second, Figure 11, is a derived model, which this author created to explain the immense knowledge and technology growth potential, that is facilitated by an appropriate research operations plan. It is the graphic model for the Concept Fusion model. It is argued that the LLNL fundamental operations plan fits the appropriate dynamic form shown in the first model. In addition, the second model describes the knowledge production potential of LLNL

The model of figure 6 shows a closed loop, cyclical control process. The figure 11 demonstrates the explosion, as in a supported exponential growth dynamic system, of knowledge and innovation production. This is examined in detail in the Concept Fusion model.

The LLNL Strategic Operations Plan (model)
For knowledge growth

The model presented next focuses on the fundamental operating requirements and mission of LLNL. The model is shown as a closed loop toward structuring a strategic plan. This is not intended as a replacement of the ESOD sketches. This model is intended to be an interpretation of the LLNL systemic operations. For this model, Vensim-Molecules software is used. The Windows form of screen is designed to be active for all of the modeling components. Computation devices are largely automatic, when desired. The Vensim-Molecules software is generally designed to analyze a closed loop systemic model for operations research. The concept that an organization can represent a closed loop is suggested by

Von Betalanffy, Lud_wig (1968). The Vensim-Molecules system helps provide rapid

adaptation and analysis of the Von Bertalanffy closed loop analysis principle. A closed loop system is one wherein the output information from a system automatically feeds back and varies the input to the original data. This in turn effects the output. The closed loop model can be draw for feed back control (information) to the original output. A closed loop definition here is that defined for control engineering and control theory. This closed loop definition for control theory is similar to the definition of a cybernetic system, except that the Vensim-Molecules software allows a broad expansion in the control factors of a closed loop system. Multiple variable systems can be shown in graphic form.

The LLNL matrix management and aggregate research activities form a systemic operation, which can be described in partial differential and vector calculus notation. This is due to the complex and interacting variables within the development of the operations. In addition, control theory for closed and open loop computation represents simultaneous differential systems. A text on control engineering is offered by Bateson, Robert N. (1993). In addition, due to the complex interaction of human characteristics, a complex domain, the organizational processes represent stochastic (probabilistic) differential systems. See also;

Kloeden and Platen (1992), Basar and Haurei (1994), Dresher (1981), Kosco (1992), and Churchland and Sejnowski (1992). Further details for this definition are to be found in the literature review of Chapter 2. Von Neumann and Morganstem (1947/1972) define the variability of human action by examining economic utility and subtle theories of measure.

Kloeden and Platen (1992) demonstrate the existence and diverse applicability of computation systems in the stochastic differential domain, and show that this is applicable to theories of psychology as well as physics processes. Basar and Haurie (1994) demonstrate that management issues can be experimentally examined using dynamic (differential) game analysis. More details for the mathematical relationships are delineated in the Concept Fusion section.

They are represented in linguistic definition rather than by a presentation of mathematical models. A very brief discussion is provided of Vector Calculus, Partial Differential systems and the connected theoretical Quantum Mechanics. This is to form a pictorial understanding of the interrelationships of the complex mathematical principles involved, with regard to the Concept Fusion model, and complex research aggregate management needs of the LLNL. In addition, the mathematical linguistic descriptions are used to provide a basis formulation that allows advanced knowledge production toward that purpose. This also provides triangular validation for the issue that standard management analysis can be improved using a more complex (Concept Fusion) analysis process.

In contrast to this, the knowledge and innovation production demonstrates an open system dynamic. This is analogous with the reinforcing systems of computer science, which demonstrate exponential growth. This concept is delineated in the Concept Fusion model.

The Strategic Operations Plan Model demonstrates that the engineering sciences operations planning that is needed can be complicated, yet follow a closed loop plan toward development of the Engineering Strategic Plan (ESP). This closed loop can then allow development of the LLNL Strategic Plan LLNLSP).

The Strategic Operations Plan model, Figure 6 is shown on page 145. The closed loop for this model progresses counter clockwise. In the upper right side, a primary box defined as Insure National Security is shown. More detailed requirements for this are found in the branching, or secondary variable box and labeled Military Strength, which is partly from nuclear weapons stewardship;

Economic Security, through general research, industrial partnerships, and energy research; Health Safety through global ecology; Education, through science and technology development.

This primary group is a summary of topics, similar to the items in the long list of activities, detailed in the LLNL report UCAR-10076-14. statement. For this These requirements form the needs basis that follow into an LLNL statement to be fulfilled, all of the variables effecting the ESOD must be taken into account. Two significant secondary boxes are shown that in turn effect the ESP, and provide the basis for the box labeled "Strategic Operations Plan" which is located at the top of the page and loop.

Continuing around the loop, loop factors are noted. The ESOD must plan for research in multiple categories. Resident information is used, which are the laboratory core skills, to allow development of the cyclic process. Also, when this occurs, the needs for increased information expand, along with development of more information and knowledge. Development of more information in turn expands the knowledge about information, and the associated sciences. This in turn accelerates technology for groups of technology categories.

Therefore, both new information and new knowledge simultaneously grow. The growth of knowledge helps the development of advancements in the categories toward the mission. This automatically improves education and the product quality. Finally, the resultant is a closed loop of support for the LLNLSP, and the final goals to insure national

security, which in turn defines and actualizes the mission.

The strategic plan model, and graph, helped provide a basis for understanding the Concept Fusion model (Cutburth, 1997), which is shown after experiment and discussion of a business analysis models. The Concept Fusion model is drawn by using the Vensim-Molecules software. However, it is not a closed loop, because it represents the actual chain reaction and virtual explosion of knowledge growth potential, which is present using the LLNL research and management methods, in a dynamic aggregated R&D plan.

One should be cognizant of the fact that LLNL Engineering Sciences Operations involve hundreds of simultaneous research programs that interrelate, to achieve the final LLNL mission. Furthermore, it is shown that the level and intensity of each research category can increase exponentially. This is suggested in the Concept Fusion model (Cutburth, 1997). Therefore, continuity is suggested, aggregation is required, and a dynamic representation using operations models, is consistent with the factors presented by the ESOD.

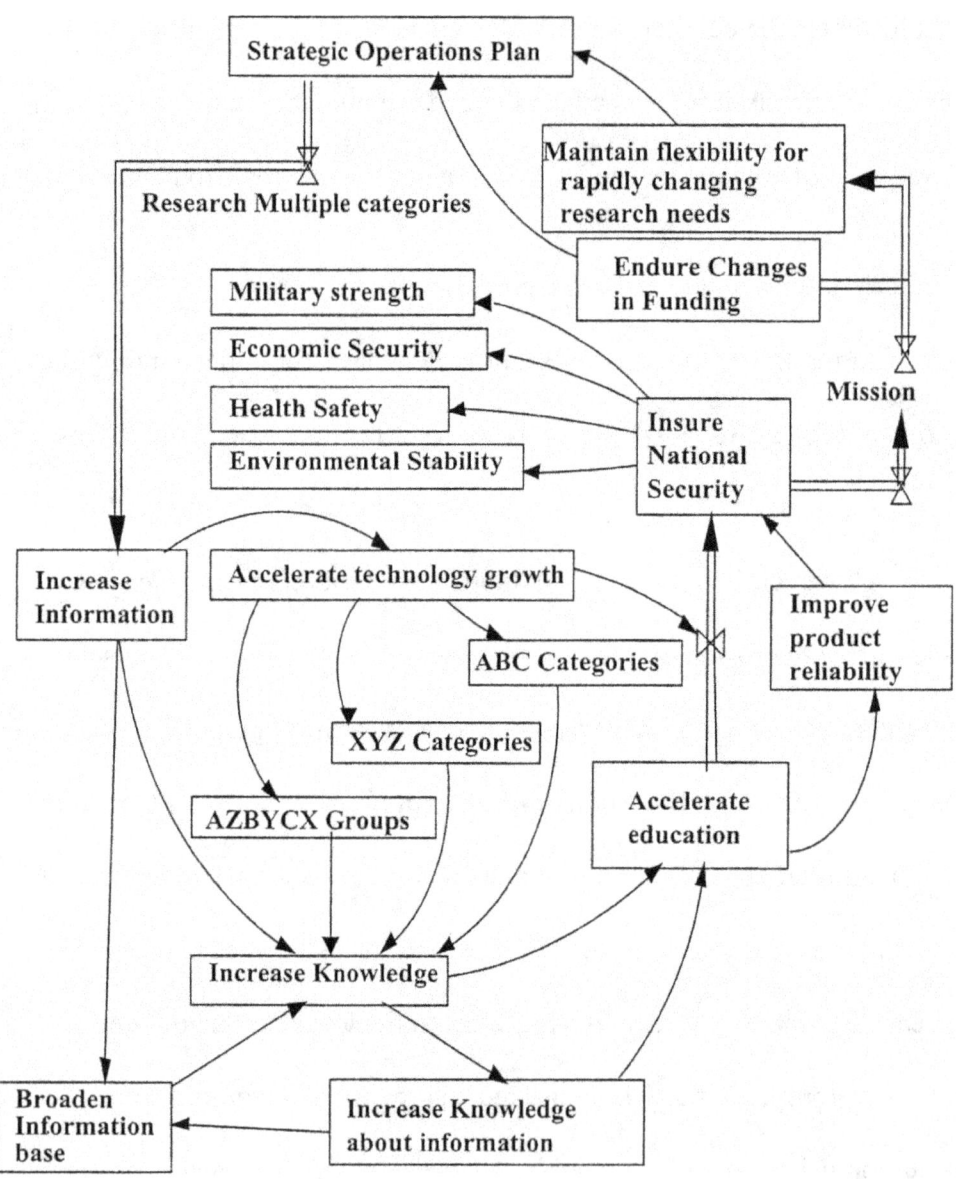

**Figure 6 The LLNL Strategic Operations Plan (model)
For knowledge growth**

195

The LLNL Simultaneous Goal Set

The Saaty (1995) Expert Choice (EC) software is used to examine the LLNL strategic goals. The software features goal analysis to assist in strategic choice analysis. A test is constructed to consider the simultaneous strategic goals of the ESOD, to achieve the LLNL ESOD Strategic Operations Plan.

The Vensim-Molecules software system allows extrapolation of subset loops within the operations analysis. The (EC) software system is designed to allow subset computation, but uses a matrix, non differentiable, and non stochastic analysis. The fundamental mathematical computation system of the EC is, therefore, contrasted to a closed loop systemic configuration, which can be drawn using the Vensim- Molecules software. The Vensim-Molecules software can represent simultaneous differential systems.

These are relevant to control engineering. Examples of differential computation and LaPlace (n.d.) Transforms are given by Bateson (1993). A partial adaptation of the EC is explored here to examine use toward the analysis of the LLNL goals. (For a complete analysis of the software, it is recommended that one contact the Expert Choice company.)

The technology offers a means to turn language into some form of computation, when this may not be understood at the time of the decision need. For a goal analysis, one starts by writing a set of strategic goal choices. A survey development analysis and method are shown which can be used to examine intricate variables with regard to the goals.

195

It is assumed that the set of goals may conflict, and that a technical analysis will help extrapolate the most desirable from the goal set-to make an expert choice. An analysis sequence is shown below which follows the general EC plan. However, one must

be cognizant of the issue here: that every ESOD goal element in this set is assumed equally important. This is a necessary factor for optimal results in meeting the LLNL strategic needs. The experimental examination below is used to test the sensitivity for change of each goal—to provide information about the LLNL aggregate requirements. It is suggested that the experiment helps validate the LLNL plan to use an aggregate of research activity, and flexible matrix management system, to achieve the interactive and simultaneous goal needs, toward actualization.

The EC software uses a windows form of data input and development. Icons provide a point and click means to select a starting point. One starts by inputting a goal set. A multiple choice of survey forms are automatically draw for possible use. For this experiment, hypothetical values are used for an experimental examination. The values are then changed to test the results. The change and subsequent test is referred to as a design scenario, or what if analysis. The EC system allows development multiple subset goal choice analysis construction. This is because some goal issues can be exceptionally complicated.

The EC/LLNL test analysis Figures 7, 8, 9, and 10. These are given as a sequence which follows the steps noted below. A singular goal is shown, which is to insure national

security. This goal is divided into the ESOD operational requirements to meet this goal ((Figure 7). This page of the computer printout sheet shows the six components of the operations goal. The LLNL goal, of national security is divided into the ESOD simultaneous goals. These are listed on the printout page 149 as ACCEDU, ACCTECH, EDNUREVE, IMPREL, INSSEC, and MAINFLEX. The initial values for the divided goal set establishes an equal distribution for each goal toward the final value of one. The initial experimental value for each then is one divided into six equal values. The intent of this experiment, as shown on the next print out sheets, is to test the results on the program if there are small changes in each of the six components. The following computer printout sheets show survey forms to collect information that may change those initial values.

For example, at the top of page 149 a hypothetical comparison is shown for relative importance of each of the six. In Figure 8, a hypothetical survey is automatically created to examine the choices of each ESOD goal element. The survey form used provides an IMPORTANCE scoring for each of the elements in comparison to the other. The results of the (hypothetical) IMPORTANCE tests shows that the goal elements would

show a spread of .029-.315. This would not meet the equal need principle for LLNL.

In Figures 9 and 10, a choice comparison is made in a hypothetical survey form, at the top of the pages, showing circled values. The example shown gives a hypothetical indication that at some point, a goal element will receive a slightly higher score. Hypothetical choices are shown which are very nearly equal. The small changes produce calculation output, as a bar graph. The results of the tests show rapid changes in the bar graph. This indicates that small changes may have an unacceptable change in the final choice group, when considering the goal issues for LLNL. (The EC offers pie charts). The resultant values can be automatically calculated for sensitivity.

The sensitivity can be evaluated for consistency. A consistency ratio, offered in the EC provides recommendation for possible element revisions, to achieve an optimal consistency ratio in the goal elements. A good consistency value is small, (around .1). The consistency is where the input demonstrates conflict in choices. A low consistency is when on one comparatively chooses in a conflicting manner. The input chosen conflicts with the choice made in an adjoining comparison. To say one does not like something, and also that one likes it in the same test, suggests conflict or inconsistency; But under varied circumstances, one may find personal conflict. Therefore the EC helps test for those issues. The EC also offers a fast method for altering the values for scenario analysis. In addition, a scenario set can be cross compared. In other words, one can make revisions in the choice comparisons to create a new scenario, in order to respond to low consistency issues.

An interpretation of the EC analysis experiment, for the ESOD needs, is that a small change in the elements from outside funding changes can easily impact the ESOD goal set. The rapid changes shown in the figures above are from small external changes. A small change can be from a small shift in revenue or congressional development priority. This suggests that it is necessary for the ESOD to maintain optimal flexibility in the operations to achieve the final goal. Furthermore, the overall planning for LLNL must have details which are far more intricate than a five choice goal set may show.

Basar and Haurie (1994) demonstrate that management choice issues represent differential systems. When considering competitive goals, and interaction of humans in competitive management systems, the analysis can be represented by stochastic differential computation methods. Kloeden and Platen (1992) suggest computation methods for stochastic differential systems, and demonstrate that they apply to issues in psychology, as well as analysis of physical processes. Thus, complex management issues and physical science analysis are augmented by this complex analysis systems.

Von Neumann and Morgenstern (1947/1972) demonstrate that human choice issues represent complex variables, and offer comparisons to the complex domain of Economic Utility. Therefore, a simple addition of choices in the EC model would not guarantee an accurate solution where stochastic differential variables are present.

The results of this EC experiment indicate that the strategic plan, which includes use of a matrix management method, and an advanced research aggregate, help retain optimal flexibility for meeting the final LLNL strategic goal.

Although the EC method provides a rapid and comprehensive means to examine a set of goals, the mass of the LLNL research aggregate must be examined by use of other methods. It is suggested that analysis of research productivity is not directly measurable through use of

standard business analysis methods. Specifically, the requirements for analysis of knowledge production and innovation in a complex research aggregate include nonlinear stochastic differential systems. The EC method is not designed for analysis of nonlinear variables. However, the EC method offers fast tools for analysis of a goal set. In addition, where there is uncertainty about how to explain the language or meaning of a choice set, the EC can help turn the words into some form of mathematical value relationship.

Specifically a set of goals, in the EC method, is represented by words. But the computation which is based on those words help demonstrate a form of mathematical linguistic found in the meaning of the words. This principle of the representation of meaning of words by mathematical values is a significant issue that is considered more deeply in the Concept Fusion model (Cutburth, 1997).

The final LLNL goal is to achieve national security through advanced research. Therefore, a focus on the research operations, knowledge growth, and innovation productivity as shown below, is consistent with the national security goal. The interaction of an advanced research aggregate, for advance productivity and analysis, is delineated under the Concept Fusion model.

The complete model is presented in the Appendix A. An outline is provided next to show its relationship to analysis of knowledge growth and strategic business planning methods. Research Planning for an Interactive Concept Regeneration —Concept Fusion

Significant research and development (R&D) interactive planning was discovered and reported in the Technology Assessment (TA) component. The LLNL programs, divisions, and thrust areas showed that projects were adopted which provide direct connectivity from program to program, division to division, thrust area to thrust area. Multiple division, program, and thrust area interactivity are predominant at LLNL. In other words, the multiple research aggregate areas provide mutual benefit for each.

This is described further as facilitating a unique Concept Fusion (CF) (Cutburth, 1997). The CF, defined by this author, is analogous to the concept of nuclear fusion energy processes. Current reports on fusion energy research define this as a method of compression of radioactive material which then drives the material to critical mass. This result is a nuclear chain reaction and an accelerated production of energy.

One principle of nuclear explosion operation is that the atoms are driven into excitation and implode or explode, allowing part of its elements to be driven into adjacent atoms which in turn become some other product, and implode or explode. A chain reaction occurs at an accelerating pace. It is claimed that the resultant energy production is greater than the energy which facilitates the fusion. The CF model is suggested to follow this form of process.

The beginning mass of knowledge is changed by a new thrust of information or focus. The stack of knowledge is driven by changes, which further result in changes to connected or adjacent knowledge areas. The changes create new knowledge and information, and a chain reaction occurs. Where the fine elements of knowledge are understood to be separable components which are initially integrated or laminated knowledge elements, a chain reaction of accelerated knowledge production can be further enhanced. The chain reaction is the production of new elements of knowledge, information, and creative concept families.

Where a nuclear explosion is limited ultimately to the amount of material used, the ideal research base knowledge expansion is limited by the current defined mission of laboratories. An expanded would be one which provides an expanding mission relevant to

the addition of nuclear fuel, were that possible in an explosion system. It is suggested here that the interactive program planning at LLNL allows development of new technology at an accelerated rate. This is because the technology can be regenerated and expanded by cyclical re-focusing and expansion of new technology developments. It is suggested that this creates the possibilities of a CF model which, when optimized, can create a broadened concept regeneration productivity. Technology acceleration is specifically optimized by the LLNL aggregated research plan. It is suggested that any research plan which follows the LLNL plan, and/or an accelerated CF principle, will experience a higher level of achievement.

A sketch of the CF process is shown in Figure 11 as it relates to a fusion energy process. Figure 11 shows a central large-scale research program aggregate. This is analogous to a large, radioactive material mass. This provides a basis for large-scale energy production. However, the CF model uses core knowledge (State Knowledge) and information for fusion and subsequent knowledge production.

The construction of this model is facilitated by use of the Vensim-Molecules operations research model software. The Strategic Operation Plan model (shown in Figure 6), showed a closed loop process formulation for actualization of the mission. The significant difference here is that the final product is a growth beyond the bounds of the initial loop. Creation of accelerated knowledge would not follow a closed loop when there is output greater than the input. A chain reaction is suggested.

The first step for the CF process is the infusion of large amounts of information into the research aggregate and State Knowledge. This can begin by the completion of a round of programmatic research. The initial infusion can begin by a new overall insight as to a mission perspective, or purpose change. The result is shown as an explosion of both knowledge and more information. A chain reaction is created. This is shown by the interaction of the knowledge and information cross components, and interaction with research aggregate components.

The ESOD increased concentration or a new information thrust allows an acceleration of the knowledge and information chain reaction, as shown in a reported thrust toward active information processing. Furthermore, the intense and diverse aggregate of research activities located at one location (LLNL) can be viewed as an interactive R&D thrusts, knowledge based—information mass, and advanced core competency mass.

Detailed arguments for the attributes of this concept are shown under the topic title of Concept Fusion (Cutburth, 1997). Comparisons to fundamental mathematical principles are described in linguistic form (not using rigorous mathematical notation). Although the mathematical linguistic form of presentation is used, the model demonstrates that complex mathematical analysis are appropriate for both nuclear processes and human knowledge production. This concept is augmented further when examining the actual functions of the human brain at the nuclear level. The CF model helps examine how complex mathematical principles relate both to knowledge growth potential and complex nuclear processes. In addition, the concepts are related to concept regeneration, as proposed by this author in a paper (Cutburth, 1981, The AFSC Dynamic Intellect Model). The model was tested by this author in the development of devices which were patented. This was accomplished while on contract at LLNL from 1982-86.

The list of patents numbers that this author created or contributed to, which report the name of Ronald W. Cutburth as inventor, are as follows:

1. 4925286
   4892283
   4811619
   4772109
   4755025
   4743763
   4703921
   4667922
   4640591

It is argued that the LLNL ESOD strategic plan for technology development provides advantageous methods for optimizing technology development. This advanced technology development can be further optimized by exploration using the principles defined for the CF model. But, to augment an understanding of the CF model, useful ideas from other authors are considered. It is suggested that the best research should include an infusion of many concepts. It is argued that past researchers have given us great ideas from which to build. It requires an aggregate of ideas and research to augment the most multiplicity of advanced thinking that can be achieved.

In the Concept Fusion model (Cutburth, 1997) shown in Figure 11, the inner box represents the state knowledge (knowledge one). Given an influx of information to the state knowledge, a chain reaction begins. The chain reaction is the simultaneous development, from extended research, of both knowledge, and raw information. The interaction of raw information, with secondary knowledge and information products, produces an explosion of products of knowledge and information.

This is examined in greater detail in the Concept Fusion book (Cutburth, 1997). The essential benefit of the finding of this expanded innovation productivity potential is that it allows the LLNL mission to be more fully understood and broadened. This is in contrast to the congressional report that the laboratories need clearer missions. The graph of the CF model follows.

158

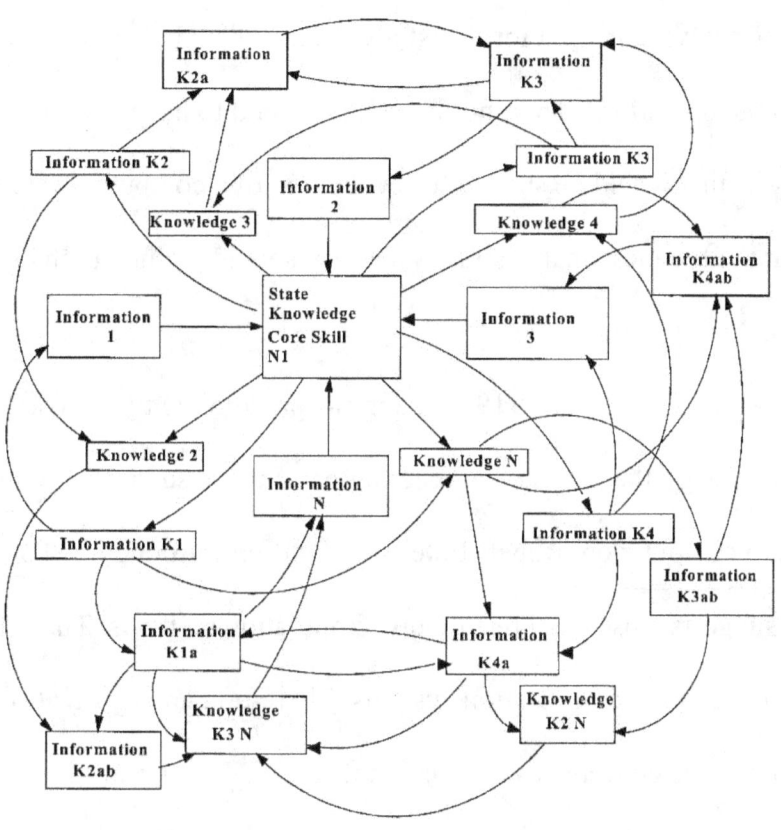

Figure 11 Concept Fusion Model

FIGURE 11   CONCEPT FUSION MODEL
Toward Concept growth
Copyright 1997 Dr. Ronald Cutburth

Systems Analysis Methods Review

A systems analysis method for this study is one which helps discover and define a System icy in the operations. This should be understood to have a broader scope than a systems analysis that would be strictly for evaluation of a computer system. But it is suggested that the broadest analysis for systemecy should include both an open and a closed loop analysis perspective.

Whereas Von Bertalanffy (1968) suggests predominantly a closed loop organizational system, the control engineering principle is suggested to allow both open and closed loop computation. Robert Bateson (1993), in Introduction To Control System Technology, suggests closed and open loop computation methods. These include simultaneous differential computation systems. Therefore, an organizational analysis, or an analysis of science concepts can be accomplished using a broader set of computation methods. In addition, the reinforcing, or open systems suggest that computation methods can be broadened, and algorithms for that purpose can grow exponentially.

Subsequently, when experimentally analyzing for the systemicy that an organization possesses, one can adapt a broad range of computation systems. But the primary point is that an aggregate of computation methods can be constructed which offer

an advanced perspective for analysis of a complex science research operation. The closed loop model drawn for analysis of the mission (Strategic

Operations Plan, Figure 6) provided evidence of the cyclical system of research to achieve the Lawrence Livermore National Laboratory (LLNL) Whereas the closed loop is

shown to be plausible for achievement of a mission, it could not represent the open loop results of an explosion of knowledge, that demonstrate a chain reaction, similar to nuclear fusion, and shown in the Concept Fusion model of Figure 11. In addition, neither the open or closed loop differential systems offer the form of data obtainable from partial differential (vectorial) computation. But, stochastic differential, and partial differential-vectorial (matrix) computations systems were shown to be applicable for experimentation as well. See Basar and Haurei (1994), Dresher (1981), and Kloeden and Platen (1992).

Standard business analysis models do not regularly include the dynamic analysis computations found in any of the above. For example, Lee (1987), Thompson and Strickland (1987), and Saaty and Alexander (1989) all have useful computation methods. But none contain the advanced mathematical formulation just previously noted. Nevertheless, the general case study methodology of the GAO, coupled with the Lacy (1992) systems analysis, helped discover the complexity of the needs for the advanced research agency. Therefore, it is in the aggregation of technologies where one finds a potential for growth in knowledge. As noted above, an aggregate was discovered in the research operations planning at LLNL.

This is relevant to the concepts used by the GAO. The interactive and connected research programs, divisions, and thrust areas can be described as a cross correlated aggregate to that defined by the GAO.

This development should be interpreted as the dynamic connectivity of concepts to describe the LLNL research planning methods and potential. Given this unique continuity, it is useful to recall the aggregate as defined by the GAO. The GAO regulation method is specifically defined in The Evaluation Synthesis (GAO/PEMD-10.1.2). This regulation teaches that a case study evaluation can be constructed by assembling an aggregate of information—to form a triangulation of validation.

It is suggested, therefore, that one examine the operations research list provided by Lacy (1992). However, one may mentally expand the meaning of the list to any criteria. For the fundamental search of the operations at LLNL, the list helped discover a set of information which describes systemicy for actualization of the mission. But also, in examining the operations, one finds complicated issues for computation of the operations.

Therefore, one could generally apply the list concepts to attempt to discover the applicable computation requirements, or the interrelationship needs to achieve research objectives. Lacy's primary area list can be considered generic. It includes: 1. Problem Definition, 2. Value System Design, 3. Functional Analysis, 4. System Synthesis, 5. Systems Analysis, 6. Decomposition, 7. Description (p. 13)

Lacy (1992) provides subsequent information lists for each of the primary areas. Part of the list for item one is shown here, because they were used to help seek useful information in this research project, under the heading of Problem Definition. The item 1. Problem Definition has a sub-categorical list.

This list is: 1. Needs Assessment. 2. Requirements 3. Constraints 4. Scenarios 5. Scope. 6. Key People 7. Degree of Need. 8. Alterable. 9. Bounds 10. Environments 11. Interactions 12. Out ofNormal Conditions, (p. 13)

The categories of inquiry present a broad potential for adaptation, which depends on the organization or need which is being studied. Some of the information found from this was shown in the interview report of the ESOD, and some is shown in the Technology Assessment. But each organization has certain characteristics that can be discovered using this form of list. For LLNL, dynamic research processes were found which show the systemicy of the operations toward development of the But also certain programmatic effects are found which can effect the overall operation and

A brief of those issues are shown along with the category of idea which is characterized by using the Lacy (1992) list. They are outlined below with regard to the LLNL operations and mission development. A few items from the Lacy category 1 Problem Definition subcategory list is used for this outline.

Lacy (1992) writes:

1. Needs assessment

For LLNL an immediate need is for advanced core skills or adjustment, and flexibility matrix management help toward optimization of those skills. In addition, each associated research category can be examined for its need to develop the other. Multiple categories of simultaneous research are shown to

be needed.

2. Requirements.

Having a national security mission component, gives a paramount or amplified need for Congressional and civilian cooperation toward development of that mission. Stable funding toward long term program development is paramount.

3.Constraints.

For LLNL, funding is not always stable. In addition, changes in congressional policy directly effect the ESOD planning needs. Federal regulation prohibits the comprehensive product development and marketing, as would be done in a corporation.

4. Scenarios.

A set of scenarios for possible changes in the planning can be used to estimate the effects of the change. This method is considered and written into the software and texts of Expert Choice (EC) software. The scenarios for examination of the goals was tested, and noted above. In addition,

Schrage (1997, p. 270) adapts Linear Programming software to allow scenario analysis . This is experimentally examined below, in analysis of employee shifting. As above, the methods demonstrate that a more complex analysis is required, because the computation systems demonstrate non-linear criteria. The issue defined by Lacy (1992) is that one must be cognizant of the possible scenario that could be optimized.

Conclusions on the Systems Analysis Method

Lacy's (1992) lists for analysis allow a broad and detailed information discovery process. The results for its application in this project have given information that supported a systemic analysis and heuristic model determination. Analysis of systemicy also included the theoretical concepts and formulation of Von Bertalanffy (1968). Therefore, the combination of the systemic search and the loop formulation provided a specific aggregate of concepts which are part of a broader aggregate of analysis possibilities.

A matrix management is found to be an essential method used at LLNL. The effectiveness is compounded by the development of an aggregate of research categories and items. In addition, the Lacy (1992) lists produced a set of information with regard to the engineering sciences operations directorate (ESOD) needs, constraints, environment, and other criteria noted above. But the significant discovery from this analysis aggregate, and as indicated in the TA and model development above, is that the LLNL method of operations offers flexibility to cope with the constraints and boundaries.

It was also discovered that the LLNL matrix and aggregated research methods simultaneously show a dynamic methodology to optimize and actualize more advanced results, in spite of some of those constraints. Complicated needs presented to LLNL are handled readily by the LLNL matrix and aggregate research methods. Planning for contraction and expansion of programs is found to be an advanced management skill area at LLNL. The general R&D programs are augmented by the strategic planning found at LLNL. However, the theory for Concept Fusion (Cutburth, 1997), and aggregation, and synthesis of concepts can be understood more fully by consideration of more analysis methods. This follows. The final product for understanding the attributes needed for an ideal research agency is shown through the mathematical connections in the Cutburth ((1997) Concept Fusion model.

Strategic Planning Perspectives

In the next few sections, strategic planning methods for LLNL are explored. Comparisons are shown from management sciences and operations research techniques. In addition, models for specific characteristics in technology management are explored. The technique used here is to examine an aggregate of analysis methods. This section is designed to explore more deeply into the connectivity and growth of knowledge found in an aggregate of research-by considering an aggregate of strategic planning methods.

It is suggested that in nearly any form of analysis or concept thinking, an aggregate-a collection of ideas- -help form a triangulation for validation of ideas. Furthermore, techniques of using combinatorial analysis, for expansion of information and knowledge elements, is shown to be plausible.

It is suggested that in the final analysis, purely mathematical concepts and their relation to language (a mathematical linguistic) provide evidence that all of knowledge growth can be accelerate through actualization of the connectivity of the mathematical principles. These are generic to a focus, and offer knowledge growth in any category. In surveying all of the texts in this section, none mention the use of combining and producing concept algorithms for computation means and subsequent knowledge growth. This concept is suggested by Cutburth (1981). For the current extension of this work, it is suggested that each of the techniques and categories of analysis found in this research project can be related to the other in direct or heuristic rational thinking. Furthermore, in most instances it is shown that there is a direct mathematical relation for one mathematical analysis method to the others. One should be cognizant of this concept, and consider the advantages of an aggregation of knowledge in research, in reviewing the techniques noted below. It is argued that this fundamental connection offers expansion of research. More importantly, a paradox in the operations at LLNL are suggested, because all of the research, which has mathematical foundation and connectivity, is adaptable to all forms of research and knowledge production in areas which have nothing to do with nuclear weapons.

A possible formulation of concept thinking through advancement of mathematical linguistics is suggested. The concept rests on the principles that reasons for language are to transmit information containing value. The interpretation of the value is in turn founded on a mathematically determined communication reasoning. It is suggested that failures in communication are not from the mathematical formulation, but are from the confusion or limited knowledge about the exact connection of the communicable mathematical conceptualizations and the value issue in consideration.

The advantages in recognition of the connectivity of thinking is the possible broadening of theoretical analysis using a group of connected and broadened concepts—wherein a more fundamental mathematical foundation in natural language may speed technology development. Recent texts on genetic programming, that use combinatorial analysis models demonstrate advancements. (See Gen,Cheng, 1996.) These are compared to the earlier Cutburth (1981) AFSC Dynamic Intellect model, in an accumulative definition, and shown in the Concept Fusion model (Cutburth, 1997). This model is examined in detail, following the considerations of strategic planning and operations analysis models shown below.

The Strategic Business Posture of LLNL

Thompson and Strickland (1987) suggest analysis methods from several large and international corporations. Strategic planning is recommended. This includes consideration of their competitive position. Methods to improve the strategic position are provided. Research is recommended to provide sufficient detail for the decision process. (p. 93) They suggests that one draw a division strategic element list. This compares the overall percentile of the structure, and strategic fit of each division, (p. 167) The financial condition is analyzed to consider the potential for change, or catastrophic developments, (p. 171) Product or system differentiation is recommended, (p. 140) To achieve this goal, a strategy may include investments to make a product more unique and useful.

LLNL has useful planning and data toward evaluation of a business strategic plan that compare to the Thompson and Strickland (1987) methods. Similar language and concepts are found in the LLNL and ESOD reports. The ESOD graphs in (Figures 3, 4, and 5) show strategic planning that includes cohesive use with the matrix management methods. In the 1996 ESOD report a re-examination of the Engineering Strategic Plan (ESP) is reported. The report provides a survey of major funding, personnel and program positions and changes. A brief quote provides a direct comparison to the idea noted by Thompson & Strickland (1987), to examine system differentiation.

The ESOD writes: In formulating our strategic plan, we re-examined the way our technology thrust areas (primary areas in which we focus our resources) are structured so we can develop technologies that are unique (otherwise why choose us?), the best (otherwise they do not belong to a national laboratory), and relevant (meaning that an LLNL program has a need for them), (p. 7)

The singular report both shows a strategic planning analysis for 1995-6, and differentiation planning by the ESOD. The ESOD sketches reported above show a strategic planning methodology for shifting focus, divestiture and area optimization. Therefore the combined ESOD analysis policy and planning parallels good business practice, as defined by Thompson and Strickland. (1987, pp. 139-143)

The charts show the Full Time Employee (FTE) shifting for the primary programs and divisions, for both 1995 and 1996. Significant changes were accomplished at LLNL.

A heuristic generalization for the Engineering Sciences Operations division is used by means of an average cost for an FTE. Therefore, the chart defines FTEs as the cost factor. This definition is also useful for showing the matrix management shifting, to meet the yearly programmatic focus needs.

Figure 14 is a replication of the chart in the ESOD 1996 report. This corresponds with the category shifts shown in Figure 12 and 13. Figure 14 is defined according to the major programs. Significant FTE shifts are indicated. In one year, 1995-1996, a 9.2% drop in work force for the engineering operations division is reported. A 7% drop was due

to a voluntary separation program, (p. 6) Now placed at 2,100 FTEs, 1,800 are in the matrix management configuration, (p. 5)

The overall shifting by program, and division areas included approximately 26% of the FTE's. The shift of more than 400 FTEs is demonstrated in figure 16. In examining the chart, one sees a somewhat proportional shifting. Large programs demonstrate large shifts, and small demonstrate small shifts. Consistency is shown for retaining the programs, but focuses are on changing program needs.

Overhead cost reductions are reported. A reduction in overall facilities was accomplished. A drop from 880,000 square feet to 800,000 was actualized in the 1995-95 time frame. The ESOD plan was accomplished to reduce overhead costs, (p. 8)

Figures 12 and 13 show the FTE shift from 1996-1997 of division to division for the Electrical Engineering (EE) and Mechanical Engineering (ME) technology categories. Although only EE and ME are shown, other engineering categories, such as electronics engineering, software engineering, manufacturing and materials engineering are generally grouped under those categories.

The division acronyms as in DSED, and MMED etc. correspond with, and are listed in, the Technology Assessment (TA) component. All of the FTE's shown in Figures 12 and 13 have distributed work among the programmatic areas shown in Figure 14. The combination of FTE shifts shown in the division and programmatic grouping demonstrate a rapid flux of change.

Significant division and thrust area interactivity was discovered through the TA. Those data are shown as MIXDIV (mixed division activities for multiple product use), and MULTICAT (Multiple categories of engineering required for the project). Other data are listed in the TA, but that information is not needed for this component area.

The ESOD revised the overall division and thrust area structure in 1996. The thrust areas are matrixed operations groups that function through the group of divisions. The strategic change is that the thrust area leaders now report to the entire engineering organization through the ESOD. The ESOD reports that: "The aims of this reorganization were: 'to emphasize the thrust areas' multidirectorate involvement; to underscore engineering's accountability to the Laboratory as a whole; and forge a better amalgamation of mechanical and electronics technologies in our key areas of research interest" (1996 Engineering Anual summary, p. 8).

The structural change would suggest an expansion of the matrix concept. This would suggest that more rapid program shifting are possible. In addition, high velocity changes in technology development may be responded to more rapidly. The thrust area directorate restructuring coincides with a focus on information engineering restructuring, (p. 7) Information exchange would be enhanced by the coupled changes. An optimized matrix system is suggested. Key changes are listed below for use in this analysis.

1. There was a reduction of work force of 9% in one year, to 2,100 FTEs.

2. There was an overall shift of approximately 26% of the FTEs in programs and divisions There was a creation of a ninth thrust area. This represents a possible Split personnel into nine groups. An average split would be from 262.5 down to 200 per thrust area.

Conclusions on the 1996 ESOD Report

An amalgamation of the primary electrical and mechanical engineering divisions through a structural change for management of the LLNL thrust areas, can be compared to engineering society activities. It is consistent with research activities sponsored by the American Society of Mechanical Engineers (ASME) and the Institute of Electronic and Electrical Engineers (IEEE). The journal report in the literature review of this project examines the new journal sponsored by joint activities of the ASME and IEEE.

The new journal is called Mechatronics. This resembles the "amalgamation" efforts reported by the ESOD of the EE and ME divisions. The ESOD structural change is consistent with research topics found in the journals from the Society of Photo-optical and Instrumentation Engineers (SPIE). Articles found in their journals and conference reports refer to considerable overlapping technology areas. For example, The Third International conference on Intelligent Materials, SPIE proceedings Vol. 2779, shows articles with the following diverse topics. "Intelligent materials and systems as a basis for innovative technologies in transportation vehicles, p. 16",

The same conference reports a diverse topic that is connected under the heading of smart materials. For example, "Biomedical materials and Structures", (p. 28) Other interrelated, but simultaneously diverse science topics are noted in the Concept Fusion section.

The Technology Assessment (TA) showed many programs and projects that use multiple categories (MULTICAT) of engineering activities to produce products. The ESOD report shows that interactive development programs existed before the new ESOD changes. However, texts recommend optimization of policies and operations planning. The ESOD changes appear to be an actualization and optimization of the historically planned LLNL interactivity.

Notes: Operations Research and Management Sciences Methods

One finds consistency in needs to overlap research areas in science topics, but also in the analysis of the operations. A primary example is found in the newly combined organizations of the Operations Research Society of America (ORSA), and The Institute of Management Sciences (ASMS). The new organization is called Operations Research Management Sciences (INFORMS).

It is possible to correlate this amalgamation (using the word broached by the ESOD) to that found in research journals. The view of this author is that the fundamental combination potential rests is fundamental mathematical principles. This is largely due to the fact that mathematical principles are fundamentally generic, and applicable to any concept. It is suggested that the development of separate organizations, prior to the integration of INFORMS is due in part to a separation of humans who are thinking in similar concepts, but are adapting their concepts in separate locations and applications. This allows expanded application, but is not necessarily the most productive optimization of concepts.

It is argued that a fundamental mathematical linguistic exists among all humans. However, a collected optimization of the principles are separated by social choices, social confusion, cultures, and history. Furthermore, ideas evolve to fit application needs, but the basis is a mathematical formulation of some form. Examples are found in the newly discovered possibility to "amalgamate" electrical and mechanical engineering research at LLNL.

Fundamental concepts for combining ideas and concepts are explored further under the heading of Concept Fusion. For this section, a concentration toward comparison of operations analysis and management sciences methods, and the application for the LLNL-ESOD operations are explored.

A comparison of three texts can provide a collected comparison of the Lawrence Livermore National Laboratory (LLNL), engineering sciences operations (ESO). Although other examples exist, three texts are chosen here. They are:

1. Lee, Sang M. (1987). Management Sciences. New York: Dryden

2. Schrage, Linus, (1997). Optimization modeling with LINDO. Albany: Duxbury (ITP). (LINDO is a software system)

3. Wilson, (1996). Operations

   Research. Pacific Grove: Brooks/Cole. (Uses the LINDO software system)

A comparison of the texts indicates why the INFORMS was formed. The topics of study use essentially the same mathematical principles, but the application approach can vary. For example; A fundamental method of analysis depends on Linear Programming (LP). Matrix computations of linear programming are a significant portion of the teachings. Each of text 1, 2, and 3 begins with the matrix and LP foundation.

The text teaching of matrix-LP analysis systems for optimization, operations research, and management sciences provides a unique fit to the matrix management system at LLNL. The ESOD shift needs are for reasons to optimize in a strategic aggregate domain.

A mathematical correspondence of the LLNL matrix management method and the linear programming concepts is suggested. Except that the processes at LLNL represents complex stochastic differential computation need, as shown above. Therefore the computations should be understood to demonstrate nonlinear variables. This is demonstrated in the topics presented in the literature review of chapter two. This is delineated by

Von Neumann and Morganstem (1947/1972), Basar and Haurie (1994), and

Hillier and Lieberman (1990). Kosco (1992), Dresher (1981), and Kloeden and Platen (1992). ESOD Strategic Variables, Differentiable Game Analysis, and Vector Calculus The texts by Lee, Schrage, and Wilson progress into analysis of strategy. Where complex variables are confronted, stochastic (probabilistic) methods are mentioned. Schrage provides a note about stochastic issues in game strategies. "Where actual personal issues are considered, even a two strategy game is stochastic", (p. 253) Thus, in human competition, complex interrelationships, or adversarial conditions exist, stochasticity exists, even with regard to the encounters of two persons.

Schrage (1997) points on decision issues must be considered with regard to the LLNL ESO decision situations. The ESOD issues include hundreds of possible issues and choices, as shown in the brief list of project and programs in the TA.

The matrix management system offers a broad set of choices, but also a broadened flexibility The mathematical game theory analysis method offers means for decision analysis. However, as Shrage advises, complex variables, differential and stochastic processes may be present. Therefore analysis of complex research science systems are demonstrative of the magnitude of complex computation that this suggest. To narrow a mathematical strategy game analysis, a two person, two strategy game, offers consideration of a small group of choices (two for each opponent in a game). Schrage (1979) provides an example of a two person, three choice game strategy. A matrix is created for an analysis of conflicting issues (or strategies, p. 255).

The game strategy methods provide an alternative for analysis of choices for a player, or decision maker. Whereas the systems offer a statistical method for analysis of choices, the game analysis methods offer direct computation methods for analysis of odds (The oddment). A significant difference exists for game strategic analysis, compared to a goal analysis grouping. .

The method of game theoretic analysis is expanded into more complicated games by Wilson (1996). A more complete definition of advanced mathematical game theory is provided by Von Neumann and Morganstem (1947/1972). They examine the theory of measure and Economic Utility. Mathematic game theory is offered by Williams (1954/1982). Continuous (differentiable) mathematical game theory is offered by Dresher (1981), Von Neumann and Morganstem (1947/1972), and Basar and Haurie (1994).

In the basic methods, a matrix is drawn wherein a value is placed in each box. A matrix of boxes is shown in Figure 15. A key strategy is to eliminate some of the possible strategies in order to simplify the game. This is difficult when stochastic differentiable games of a large order are involved. Because each of the values in each box may represent a differentiable variable. In addition, they are stochastic when uncertainty about the starting value exists. Therefore, although a matrix can be the starting graphic picture, the computations are complex. An issue may exist, where computations can become more complicated. This is when the variables are dependent upon differentiable variables found in any of the other boxes.

To consider this matrix computation outline to the needs of LLNL, one must attribute programs, projects, employees, funding, or even technical science areas to the boxes. This effort in its self represents complicated set of variables. Because variability in human characteristics, resident core skill, funding stability, and physical sciences. However, the computation systems exist for experimentally examining this form of analysis. Honig (1986) and Gaynor (1996) suggest layers of matrix systems computation, for examining factors which interrelate. However, they do not define or delineate the compounded variables that are demonstrated at LLNL. Saaty (1995, Expert Choice) allows development of layers for goal analysis, but the numerical systems are linear, and determinate on singular goals. To develop a simplified strategy, a player will attempt to find groups of strategies and patterns of playing which will have little (or ideal) value for victory. The player may alternate the activity among group of choices in a form of a pattern known as the oddment (playing the odds). See any and all of the game strategy books noted above for this common definition of oddment. It is argued that this form of adaptation of strategy and odds playing, is relative to using cyclical or variable project funding strategies A significant difference for this point is that a matrix computation that is linear produces set of linear output numbers. Therefore, linear computation methods will allow error for areas where compounded variables exist.

There are other complex variables which are noted in the Concept Fusion model (Cutburth, 1997). However, one should be cognizant of the fact that the existence of complex variables, as suggested by this author, is a representation of the expanded possibilities in knowledge growth. Specifically, when there are expanded variables and possibilities, greater creativity is possible. In the example of Figure 15, a two player competitive matrix game is shown. Each player has an opposing strategy, which is simply the pattern from the side which is represented at 90 degrees from the other player. The one player plays the odds in the values for the columns and the other plays the odds in the values for the rows. A significant concept analogy is added to this game understanding. This is that the players, in finding ideal strategies will select a group which are referred to as channels. These are later referred to in concepts from theoretical nuclear physics. This is examined for that connectivity because when analyzing the ideal channels, one must be cognizant of the issue and predominant need to help knowledge grow in all of the research activities. Specifically, knowledge growth is a final objective, no matter what

form of project name or definition is placed on the matrix, where research is concerned.

In other words, a key concept is presented. This is that knowledge growth, and project planning can be for the overall project choices, the funding, or provide a

230

definition of the complexities for knowledge growth of an individual. Kosko (1994)

suggests that neural networks represent a matrix computation.

One type of learning is called differential competitive learning. He provided

connection of many forms of science categories through this method. The point drawn for

this work is that the addition of the new work of Kosko helps connect concepts.

In fact this is required for expanding knowledge in advanced research. This is

delineated further in the Concept Fusion book (Cutburth, 1997).Inasmuch as Kosko

(1992) uses the same principles of the authors on the mathematical methods shown

herein, one must understand that the game theory chart shown in Figure 15 must be

considered generic to a specific focus—it provides methods of learning, and is suggested

to corresponds with the matrix management systems followed at LLNL sufficiently to

allow experimentation for that purpose.

230

232

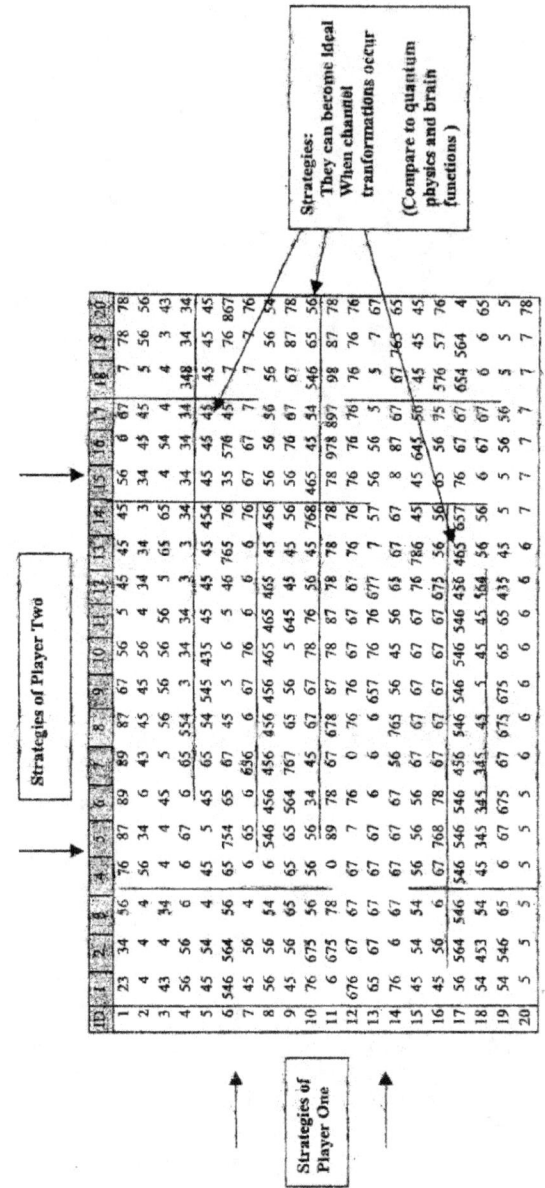

Figure 15 Mathematical Game Theory example chart

FIGURE 15
GAME THEORY CHART WITH TWO PLAYER FOR LLNL STATEGIES.

232

It is suggested that the LLNL matrix management system offers a direct means for advanced game and research aggregate analysis methods. Except that large scale computer systems are required to manipulate and control all of the complex variables. However, complex issues, as suggested in the issues of channels, and subsequent innovation development, could demonstrate a computation set as being chaotic. This is due to the interconnected issues which may represent a nuclear physics process, as well as the variables in learning processes.

Lee (1987) does not provide a definition of stochastic or differentiable games. However, Lee provides chapter for analysis of complex variables in decision analysis, using Markov (probabilistic) processes, (p. 685) A demonstration is given of special cases

of Markov analysis. Lee suggests development of a matrix of transition probabilities, (p. 696) Large probabilistic matrix systems are experimentally evaluated by partitioning methods. (Partitioning is directly related to the ideas of finding channels in a game matrix.)

In other words, given a large group of decision issues, one can use matrix and partitioning methods to group the number of choices. However, Lee (1987) and Schrage (1997) make clear definition that decision issues can be probabilistic (represent uncertainty). Lee (1987), Schrage (1997) and Wilson (1996) use matrix and linear programming methods for decision analysis.

This suggests that the ESO matrix management methods can be experimentally examined by this means. But addition mathematical modeling methods are needed. Von Neumann and Morganstem (1947/1972) show that, for economic analysis, a zero-sum game does not allow inclusion of all the variables. A zero-sum game is where a specific "pot of money" or values are drawn from by each player until one holds the whole pot of money or values, (p. 504) In addition, an economic analysis using game methods can include many players. This is referred to as the   2- person game. The combined large scale n- person, and infinite choices for non zero-sum economic Issues produces a system of calculations using simultaneous differential systems. Differential Operator notation and application are suggested

The matrix for the ESO can include interaction of nine divisions, and nine interacting thrust areas. Within this structure, there are 1,800 matrixed FTE's reported. This value is considered by this author, as applicable to the principles explored in the Concept Fusion model (Cutburth, 1997).

The LLNL can be involved in more than 150 significant individual Research and Development (R&D) projects in one year. The technical support for these ranges from a few to hundreds of FTE's. Although the large programs have hundreds of FTE's, the programs are dependent on smaller research and development projects.

The ESOD revision, and further actualization of the matrix management system, suggests evaluation of 2,100 individuals is possible using the CF model. An objective of the CF model is to add mathematical concept generalization, to augment improvement of the innovation productivity analysis of a research agency. In considering innovation productivity, the directors must be included. But each employee does not represent an integer value, because of the core skill and knowledge focus of each represents a variable. See Basar and Haurie (1994), Schrage (1997), Von Neumann and Morganstem (1968), and Dresher (1981).

In addition to the stochasticity and simultaneous differential systems present, the multiplicity of dependent variables are suggested, because of the interactive needs of one program to another. Therefore, computation parameters are present suggesting adaptability of Vector Calculus. For example: given three interactive and dependent product developments, where one is effected by the other two, a Vector Space (cubed space) method or vector computation is suggested. See Marsden and Tromba (1988)

Vector Calculus, for detailed partial differential computation methods in Vector Spaces, Partial differential computation, which is part of the vector calculus computation systems, is the step- wise computation of each of the differentiable elements, in each of its existent variables, which affect the items of interest.

The items, in this case a project, are to be envisioned in a current state of development as if it were a ball in that box undergoing multiple encounters and internal changes. But the ball has three dimensions to move, X, Y, and Z. To move in any direction, a change must take place.

This necessitates computation of all of the variables, X, Y, and Z. But interaction of many projects requires experimental examination of all of the variables that are present. Simultaneous, and step wise computation of hundreds of balls in a box (or complex science projects), and thousands of interacting variables for hundreds of interacting projects exist in a complex interactive research environment. This suggests large main frame computer computation needs.

Given the existence of stochasticity (forms of probability), that is to say that it is unknown where each of the hundreds of balls in the box will be at any given time, or their condition and size, at the beginning of the computation cycle. This relates to Quantum Physics and Quantum Mechanics. (See Eisberg and Resnick 1974, and Schiff ,1968.) Considering this ream of computation systems, one may experimentally examine the effects using generic mathematical computation systems. For example, see Eisburg and Resnick's book, (1974). Quantum Mechanics of Atoms, Molecules, Solids Nuclei, and Particles, to see the somewhat generic application of the mathematical models.

Some of the variables for the balls in the box with regard to the projects can be:

Ball mass = Project size

Ball spin and direction = project to project relationship affects.

Ball bounded states = project bounded dependencies.

Ball group states = projects groups which are a set, which impact project sets, as in atoms with an electron. The complete treatment of this theoretical area is outside of the scope of this project. But the actual variables in the complex research, as in nuclear particle analysis, include a great many more variables than can be briefly outlined here. It should be seen that the matrix or vector calculus and other associated

physics models can apply to any object that is envisioned in the box. The singular or group of balls in the box which interact, may be research projects, molecules, or atoms. One singular atom may have its orbiting electrons, and the computation system is based on the frame of reference, which is usually a box, referred to as Euclidean space, which is derived from the mathematician Euclid. Therefore, the matrix space becomes a cubed space which allows more detailed analysis of the compounded variables in a complex research system. The adaptation of Differential Operator notation is used to simplify the exposition of the complicated mathematical computation systems.

Examples of simplified notation are given by Birdsall and Langdon

(1985). Eisberg and Resnick (1974, and Schiff (1968). The principles of this are

examined in the Concept Fusion section below. However, two fundamental points are

drawn here. One is that all of the computation systems relate in some way through the

concepts discovered in the human mind. The second is that the assembly of all of these

concepts allows a more broad and fruitful potential for knowledge growth, and should be

the basis for an ideal research agency for the United States. The knowledge basis and

connections are examined using the Concept Fusion model.

Notes for Optimization Model Applications.

Schrage (1997) provides operations analysis methods, and demonstrates the

application of the LINDO (linear programming) software. The software system is used

here for analysis of the personnel

shifts that regularly occur at LLNL. The ftill time employee (FTE), for this analysis,

is averaged as units. In the LINDO model they are considered integer values, because the

system manipulates linear variables.

Considerably more variables exist, however. A more complete detailed analysis
would

require a large scale computer system. But the LINDO and Schrage analysis is used here

to help experimentally examine the overall program FTE shifting.

One should be cognizant of the fact that LLNL has created large scale software systems for analysis of large scale complex variables for nuclear physics processes, structural systems, genetic models, chemical processes, global warfare analysis (and others). For example: DYNA , is a finite element analysis system developed by LLNL. See UCRL-ID-112607. Other large computer (mainframe systems) are SISAL a modular mathematical modeling system, and EIGER is an electromagnetic field computation system. See SISAL report M-146, and EIGER report in UCRL-53 868-95.

As suggested above, large-scale mathematical game analysis can be directly applied for experimentation of the LLNL matrix management system. The optimization modeling suggested by Schrage (1997) is experimentally used for interpretation of the effects of the FTE shifts. But it is suggested that the software methods can be improved upon in an advanced research environment, as is found at LLNL. In the Concept Fusion model shown below, the actualization of software algorithm production is shown as a dynamic process, which is an emulation of the real activity and capability of the human brain.

The LINDO software system offers useful analysis methods. However, it is suggested that one consider the system as a method which can become transformed for expanded adaptation. Furthermore, one should visualize that the FTE shifting mentioned here may be a representation of the human mental transformation processes that is attainable by a flexible management plan. In other words, the FTE shifting analyzed here is more than a budget shifting issue.

Shifting optimization should represents an education growth process for the FTE participants. This was partly described by the ESOD as development of the LLNL core technology areas. Benefits for this are examined further in the Cutburth (1997) Concept Fusion model.

Schrage (1997) begins by describing linear programming techniques, (p. 1) This is Expanded into discussions about sensitivity analysis. Sensitivity analysis was noted above as part of the Expert Choice software system. By definition, a sensitivity analysis is a re-examination of the calculation output values to test whether they seem to balance against one another. Given a set of variables, one should evaluate how easily the data can effect the interpretation. If changes can occur rapidly or easily, then the data exhibits sensitivity. The Expert Choice model is designed to examine the sensitivity for a decision,

but the questionnaires also allow an expansion and test of emotional sensitivity about the decision, sense word choices are the basis for the analysis. Although this is not the rationale for the design focus of the software, it allows consideration of issues that specific linear programming analysis may not encompass.

Furthermore, the Von Neumann and Morganstem (1947/1972) game theoretic methods are designed to allow multiple choices that are driven by human emotion choice issues. But the choices are given as a group of strategies, against a competitor's group of choices or strategies.

Schrage (1997) uses the LINDO software to examine several sets of data which exhibit multiple variables, and potential for change. To set up a set of variables for analysis, one may use a matrix. Schrage mentions game theory adaptations in a brief chapter (pp. 250-257). However, stochasticity of human action is covered under a separate chapter. To analyze variances caused by stochasticity, a scenario method is used. The scenario is the creation of a second matrix, containing changed values in the input. Then the results from the changes are then compared (pp. 224-228).

One should be cognizant of the issue that the Schrage (1997) use of a matrix scenario does not suggest a formal mathematical treatment of stochastic or differential systems. Where human action is involved, especially in a large group, and where hundreds of projects are involved, stochastic differential analysis considerations are suggested. Examples about the computation concepts are provided by

Kloeden and Platen (1992). But, where matrix management methods are used, a matrix computation offers a mathematical and linguistic generalization, which demonstrate that a mathematical connectivity exists in varied categories of analysis.

The scale required for a direct adaptation to the LLNL programs is outside the scope of this work. However, one can visualize that the inclusion of the multiple variable potentials, defining stochastic processes, can demonstrate a broadened human decision potential. Specifically, multiple variables, it is broader potential for suggested, describe a broader potential for productivity of each FTE, as well program shift optimization for the ESOD.

A comparison of LLNL shift data is experimentally examined from Figure 12, 13, and 14 (shown above). A hypothetical multiplier (coefficient) is used to compare the advantages of one shift to another toward FTE program realignment. An actual human growth factor analysis would be extremely complicated, because one can estimate that one person will find a higher knowledge growth rate by a shift to a completely diverse area, but another will find the greater growth from a new application in a similar area. Education processes should be considered stochastic. It is suggested by this author that exposure to multiple programs provide

advantages for expanded productivity potential. This is delineated in the Concept Fusion model. The LLNL shifting considered is from the 1996 LLNL report noted above, and shown in (Figure 14, p. 177). The shifting of FTEs in Figure 14 was caused largely by Congressional fund shifting (see UCRL-ID-127001). This is experimentally compared to statistical data values from the Technology Assessment component, shown in (Table1).

Therefore, computations were constructed from the TA data to give hypothetical values for computation coefficients to compare with the LINDO software test. But also the actual shifting from Congressional funding shifts is computed to provide a similar computation coefficient. The actual shifting from Congressional funding shifts is shown to be similar in the coefficient value to the specific spread in the coefficient shown in the TA hypothetical values. However, one must be cognizant of the fact that none of the projects in the TA demonstrate that they should be reduced or eliminated. Nevertheless, the funding shifts represent a gross change that can eliminate those projects whether they should be or not.

To construct the coefficients for computation, a composite of information is used from the TA and from the ESOD 1997 report. From the TA, data from Table 1 is used. This table provides tabulation of all of the Base Line Scores and project application data. The mean score for all of the projects, and shown on Table 1 was computed to be bulk mean value of 57.16. This was the mean value of all the base line scores of all of the projects surveyed, based on a hypothetical global value average of 50. Although this can be experimentally fluctuated, it is used hear as a central computation number. From this value, the program group scores are assessed. The group showing the lowest and highest mean values are 50.00 and 75.00 respectively. This provides a bulk maximum and minimum with relation to the laboratory bulk mean. The set could be shifted for other complex analysis methods. However, for this experiment, these are the values used.

The spread of the high and low groups is partly due to the issue of the size of the overall operations programs, and projects found at LLNL. But all of the scoring means provides useful information for FTE shift considerations. In addition to the bulk scores, a value ratio for project desirability is computed bellow. The information for this is drawn from TABLE 1. The amount of yess that are shown for a project, show the diverse applications for it. Therefor, a count of the yess for each of the projects is tabulated

The score provides that one yes =1. Except that a project that shows a patent received yes=2. score of a possible 1-9, the mean was computed to be 4.992. To provide a high and low of this, the standard deviation was computed. This is; std dev = 1.681. Therefore, the desirability of the project may be from a high of 6.672 to a low of 3.311: for computation purposes. In reexamining the statistic for desirability, one should be cognizant that a mean laboratory score of nearly 5 yess means that there are 5 significant areas where a project may have repeated usability. In other words, the LLNL standard for

desirability shows that there are likely 5 significant reasons to develop the project—as opposed to one, which would perhaps get a project started. But the low computed value for LLNL is at least 3 good reasons to start a project. One compares this to product marketability.

The summary of values, and computation provides a wide spread for the

Experimental analysis shown below. The collected ratio values and basis formula for the

coefficients are:
Laboratory mean score = 57.16
Program or group score mean minimum = 50.00
Program or group score mean maximum = 75.00

Laboratory mean of project desirability = 4.992 yes for project duplicity.
The desirability minimum and maximum= 3.311 minimum to 6. 672 maximum yess
The composite formula for deriving the minimum and maximum coefficient value that
can be experimentally used on a project or program is: FTE, (n) ((X / 57.16) (Y / 4.992))
where X= bulk score statistic min or max, and Y = the bulk max or min desirability
statistic. Computing for the combined maximum and minimum values gives a bulk
coefficient value minimum of 0.67 and a maximum of 1.75. This wide spread suggests
that project shifting flexibility for efficiency and desirability might be needed. But, one
must be cognizant of the fact that the low value of 0.67 does not suggest that a program
has low performance or marketability. This is because, none of the items in the survey
that comprise the data for this coefficient, show specific reasons to be eliminated. In
addition, for this experiment, it is shown that, though a score for a project is high, it could
have minimum desirability, and drive the coefficient down.

The issue of desirability may have a greater weighted value than can be shown for this bulk calculation. The combination of low desirability and low score can suggest project adjustments are needed. Schrage (1997, LINDO) software module. It is not assumed to be the most exact coefficient computation basis, because many more variables

exist than can be shown by this test method. But the values do correspond to the actual shifting from Congressional funding shifts. In all cases, the choices of the meaning are as broad as this coefficient value computation could produce. None suggest a connection to the innovation potential or differential product value.

Nevertheless, a coefficient value is used in the Liner Programming software adaptation that is experimentally used below. The management of complex science research operations is shown to exhibit stochastic differential computation needs. This is suggested by Schrage (1997), but not used in the linear computation software of LINDO. In addition, Basare and Haurie (1994), Von Neumann and Morganstem (1947/1972), and Dresher (1981) demonstrate that complex operations and all human action issues, as in economics, demonstrate stochastic differential computation needs. This analysis and issue are delineated below in the Concept Fusion model outline. However, bulk computation methods, as shown by Saaty (1995) and Schrage (1997) offer perhaps temporary issue consideration.

The score coefficients shown above suggests that an effort to increase the market potential, by increasing the application duplicity can be optimized to help offset a less than desirable intermediate bulk productivity score. In fact, a large overlap and interactivity of projects is shown in the TA. This is why the mean laboratory value for YES's is nearly 5. Therefore, one may interpret that the high coefficient spread is due in part to the LLNL planned increase in project duplicity. Nevertheless, this spread can produce an active need for FTE shifting which is optimized through the LLNL matrix management plan. In addition, a secondary program shifting potential, as in thrust areas, allows a more rapid adaptation to a focus area. Therefore the actual shifting would exhibit

a more volatile relationship of values. The volatility suggests the complex dynamic variables that quantum mechanics is designed to account for. The Schrage (1997) text and LINDO software package offer several methods for considering model analysis. It is recommended that one compare the entire text for possible intricate applications. However, the melding of each of the methods would be best done on a large scale computer system. But, as stated above, the Linear programming methods of LINDO do not provide for a formal adaptation of stochastic differential systems, nor issues with regard to knowledge and innovation production. The complex science research aggregate of LLNL are characterized by those concepts and advantages.

The matrix computation methods of LINDO demonstrate the mathematical connectivity of a matrix management system to matrix mathematical computation. But, it also demonstrates that an aggregate of computation is needed. Because, linear manipulation of linear variables, as in FTE shifting, does not attribute the nonlinear stochastic differential systems that apply in the human characteristics, and with regard to innovation parameters, that exist in a complex research agency.

It is argued that, although more intricacy can be achieved, the essential approach in the LINDO software does not allow development of an understanding for the actual potential value of a complete research agency. In addition, FTE shifts required by congressional funding shifts were greater than the assessment analysis requirements and thus greater than the LNDO software could determine appropriate for actual lab needs.

CHAPTER 5

CONCLUSIONS

There are two significant areas for this chapter. The first is with regard to the general management practices found at Lawrence Livermore National Laboratory (LLNL). The discussion is with regard to list of questions, that are part of Chapter 3, on the engineering sciences operations policy. There are side issues in this research project which are outlined below. The purposes of this study, as noted in

Chapter 1 follow:

1. To evaluate the engineering sciences operations to determine the mission of national laboratories, focusing on LLNL, including analysis of the mission statement. This will be to examine the current activities which may clarify or eliminate the conflict found in government publications about the LLNL mission, and its management methods.

2. To evaluate the engineering sciences research and development operations, for determination of its In addition, this research will be with current activities and capabilities to support the LLNL regard to the possibility of an upgraded mission statement for LLNL. This will be an evaluation of the simultaneous existence of the actual, and a resultant mission.

A significant result of this study is the discovery that the LLNL mission, and its required operations to meet that mission, actually demonstrate a paradoxical, but socially beneficial, research Although nuclear weapons stockpile maintenance is a significant laboratory purpose, the research to accomplish that mission involves all categories of knowledge production through technology development which is not nuclear weapons centered. In the process of this research, certain side issues became prominent. One is an issue as to who should evaluate the actual effectiveness or efficiency of the massive and complex LLNL operations.

The Department of Energy (DOE) has charge of LLNL, through political hierarchy. Yet the DOE does not focus its attention regularly toward the purpose of optimizing all of the activities of LLNL, as a specific evolving benefactor of the United States. The interests of the DOE are divided by political priorities which are partly driven by presidents who rotate. This is because the director of the DOE currently holds a Cabinet position. Overall management of LLNL is also shared by the Department of Defense (DOD), which requires significant authority for the products which augment and insure national security. In addition, the congressionally mandated direct requirement for LLNL to cooperate with private industry and all research activities causes a direct requirement to accept cooperative leadership with those entities.

The General Accounting Agency (GAO) has specific responsibility to audit and report to congress on the effectiveness of all congressionally funded operations. In addition to the diversity of management input, the LLNL is not allowed to develop products as profit-oriented activities. Yet, it is sometimes expected that LLNL must also manage itself as a corporation. An example of that school of thought is when the DOE asked a corporate body including corporate executive to examine the LLNL efficiency.

This became the Galvin report. Galvin (1995) rested many of the LLNL management problems on the DOE. It is argued by this researcher that the issue of corporate versus government management can detract from the development of a clear The issue of the method of measure and of rationale for accounting became prominent, adding to complexity of this research project. Since the GAO had stated that the laboratories need better management and clearer missions, their approaches to case analysis were examined in depth. The GAO has evaluated operations of both the DOE and the national laboratories, including LLNL. It was noted above, from GAO reports, that the national laboratories need better management and clearer.

The GAO methods proved to be effective for general management evaluations. It was argued in that degree project report, by this researcher, that the GAO methods work well on that form of evaluation, and that the DOE did not manage the program well. It must be understood that the research project was not designed to compare or evaluate laboratories, or to evaluate the effectiveness of the DOE. It was designed to help sort out the issues about management and analysis of any research agency. The GAO methods were therefore adapted to augment this research program. Although the GAO methods, as noted above, were helpful for the concept to build an aggregate of analysis material, it was discovered that the methods were not designed to examine the technical issues required of the advanced LLNL research agency. The void to be filled was shown to be the analysis of theoretical knowledge and subsequent innovation production needs for this nation and of human kind.

This void was filled by examining the broad perspectives on operations management methods and political and legal forces. These were simultaneously cross validated through development of the Cutburth (1997) Concept Fusion model. This model demonstrates that knowledge and innovation production can be claimed to be central to the LLNL mission. It also demonstrates that standard operations management methods are not appropriate for use in management of LLNL

Systems and Operations Questions: Regarding the Engineering Sciences

Operation Policy In Chapter 3 above, a list of questions were itemized for consideration of the Lawrence Livermore National Laboratory (LLNL) engineering sciences operations. Answers to the questions are considered below, based on the item number shown in the list. The answers follow this list:

1. Do the operations follow norms found in non-government entities-industry? Can it be said that the operations parallel business ?

2. Do the operations experience catastrophic developments due to changes in government policy? Or, does the operations design allow proportionate changes? Can a change impact be compared to market changes for industry? Is LLNL able to respond to changes effectively?

3. Does the military project research planning allow technology adaptation for non-military use? Does the operation plan lend toward a clear duel use benefit?

2. Does the State of California provide financial support of the LLNL research operations?

3. Are systemic needs internally consistent—toward efficiency and creativity?

4. Do information sciences technologies and planning permeate the operations?

7.. Given a known nuclear weapons responsibility, can one find a central function which is not actually nuclear weapons oriented?

8. Do technology skill areas seem to operate at the same quality level in all activity areas?

9. Does the LLNL location appear to be appropriate? Can significant location questions be found?

10. Do the operations provide benefit to national education?

11. Can technology research areas be found which may be effectively divested or added?

In addition to the answers considered, additional information and concepts were discovered and presented in Chapters 3, 4, and 5. The information provided below can be validated when reading the entire work. But summary and conclusions are included here because they help provide an overview of key research information.

The LLNL operations at least parallel the advanced planning and best practice as indicated by Thompson and Strickland (1987), and Lee (1987) that is used by industry. This is shown to be true in several categories that are noted in Chapter 4. For example, a matrix management method is used to augment a flexible work force posture. This was shown to be required because of the mix and changeability of priorities from the DOE and the DOD, and from ongoing internal product evolution.

255

In addition, strategic planning was discovered which strongly emulates those for strategic management of large scale and international corporations, as shown by Thompson and Strickland (1987). It is shown that the planning needs are more complicated than has been historically understood by reviewers, and perhaps Congress. LLNL clearly emulates large international corporate and small business activity methods simultaneously when comparing to Thompson and Stickland. These diverse categories are found in a unique blending at LLNL.

The LLNL uniqueness was discovered to include multiple methods for virtual and direct product development. The developments include ultra minute and delicate technical products and massive integrated research facilities, which simultaneously develop many technical products and technology thresholds. Yet, all of these are integrated in one large massive operation, toward optimal R&D, as opposed to separate entities, as are often sponsored in corporate planning. distinct difference can be found in corporate divestiture and investment to that of LLNL, because LLNL does not completely control the choices for large program ventures as do corporate executives. Congress controls the overall funding choices. This creates large-scale program change potential. However, LLNL demonstrates unique skills at providing flexibility. This is accomplished through its aggregated and laminated management planning.

A contrastive issue must be considered. Although LLNL does not make the final decision on funding choices, the customer base (Congress) depends upon the LLNL developed core skill areas. LLNL has Conscientiously nurtured the advancement of core skill areas for the benefit of the United States. The core skill areas span many product and service areas. In other words, LLNL seems to exhibit a skill to maintain its marketability for insuring continued effectiveness. However, current business analysis methods are not fully useful to LLNL. Thus a direct comparison by use of those principles is insufficient.

The military project and product core skill areas at LLNL are shown to be adaptable to an aggregate of product and science advancement areas that are not for making war. These are clearly for improved national economic, benefit and physical health security of the US citizen. In other words, the LLNL projects overlap and connectivity planning demonstrates dual use benefit in its research.

The State of California provides no visible funding for LLNL research operations activities. This seems to conflict with the fact that LLNL conducts research activities which are beneficial to California. Systemic activities toward an effective operation were shown to be consistently appropriate for actualization of the LLNL mission. Specifically, advanced models were drawn which demonstrate consistency. These are the LLNL Strategic Operations Plan, and the Concept Fusion model (Cutburth, 1997).

This helped provide evidence that LLNL does not simply respond to funding. The systemic planning was shown to be designed to be generally effective in planning to respond to some funding instability. In contrast, the systemicy demonstrates growth potential for optimizing the LLNL innovation potential. Information sciences is found to be a key element of the conceptual development planning at LLNL. A thrust area includes

an aggregate of research in this area. This is designed to augment expanded knowledge growth of technology and interconnectivity development. This was found to be an integrated part of the systemic plan for the overall

Although this was current for the research discovery, it2cnco

is expected that LLNL will transform in other aspects to meet the challenges and changes

in technology. A controversy can be shown with regard to the central purpose of LLNL. Funding is toward a safe nuclear weapons stockpile stewardship. This includes the recent funding for development of the $800 million National Ignition Facility. However, as detailed by the research for this project, the LLNL could revise its mission statement, because the dual use planning from the funding helps accelerate technology which is not connected to nuclear weapons. The Technology Assessment (TA) showed connectivity to many other topics. This was demonstrative of the LLNL planned core skill development, as well as its systemic connectivity.

It is suggested that the military directive and LLNL planning, which provide multiple use from research, have been effective. But the results show that the actual value center is not toward nuclear weapons, but other components of national security. These are found to be global: health, economic stimulation and advancement, and knowledge production.

A transformation by relativistic knowledge expansion is shown. This is demonstrated in the Cutburth (1997) Concept Fusion model. The model shows the correspondence of the operations of a nuclear explosion to human knowledge growth potential, when adapting an ideal knowledge growth plan. For example, although nuclear fusion research can be for energy and weapons, a Concept Fusion (Cutburth, 1997) is the actual result which has a relativistic transformation base, and therefore is adaptable to, and is found to effect all of the other aggregated projects. A relativistic knowledge explosive growth potential is shown.

Not all of the divisions show the same level of quality and productivity. Of nine divisions found in the TA, some moderated differences were shown. The mean score for the divisions show a spread showing a low of 50.56 to a high of 61.30 compared to the empirically defined national average of 50.

However, the measure was done for experimental purposes, to examine the overall operations needs and One could expect that of nine divisions, a difference in planning effects, toward development of the normal—but not without generating improvement needs. No division appears to demonstrate performance below a high level of achievement.

It is argued that outside funding changes can effect the efficiency levels, because changes can be rapid. Rapid, and large scale, personnel shifting is shown in the data for 1995 through 1996. Nevertheless, the overall scores were not shown to be exceptionally low for any division. This is difficult to achieve, and hard to ignore.

The LLNL locations near Livermore and Tracy, California, have many valuable advantages. However, the issues with regard to the lack of funding, and large taxation and

regulations from the State of California could cause controversy for the value of these locations

.

The Concept Fusion model (Cutburth, 1997) outlined in item 7 helps demonstrate that the overall knowledge growth potential from LLNL is far in excess than what is historically seen. The knowledge production potential, based on the current programs, was mathematically demonstrated to allow an explosion of idea mutations and program expansion, that was not shown in any congressional or LLNL report. Furthermore, it is shown in the LLNL reports that LLNL provides education advancements through several cooperative research and education activities. (See the LLNL Institutional Plan: FY 1996-2001 UCAR-10076-14.)

It is argued that the ideas which are promoted historically under the topic of technology transfer should be revised. It is shown, through examining the Concept Fusion model, that all methods of knowledge definition and dissemination can be revised and accelerated. LLNL shows research planning methods which were demonstrated to be ideal for national research. In demonstrating the underlying concepts, a new era of knowledge definition on a national scale, as delineated in the Concept Fusion model, is shown to be needed.

The LLNL engineering sciences directorate demonstrated that a plan is in place which emulates large scale methods for divestiture, but for a technology development. It is shown in the TA that LLNL has parallel planning methods and concepts toward those purposes. However, it is argued that a great deal of technology development and exploitation can be augmented through a new analysis and appropriate funding. It is shown through the Concept Fusion model (Cutburth, 1997) that an innovation science can exist. When using the principles shown in the Concept Fusion model, as in combinatorial concept evaluation, one finds significant potential for expanded product development.

The Cutburth (1996) Concept Fusion model
research aggregate Finding

The research aggregate conduction provided a broad base of information about the

engineering sciences operations. This in turn produced new information about the actual

capabilities of LLNL which exhibit a broader capability than was expected prior to the

research. At the start of the research it was known that a conflict existed about the

mission of LLNL, as shown in the congressional records, noted in Chapter 1. The

research information demonstrated that LLNL conducts extremely broad and diverse

research activities. This was demonstrated to have interconnectivity at every level in the

engineering sciences operations, as shown in the Technology Assessment (TA). This

interconnectivity provided the basis for the creation of the Concept Fusion (CF) model by

this author, which is consistent with the actual advanced levels of innovation potential

demonstrated at LLNL. This model, as detailed in the Concept Fusion book (Cutburth,

1997) suggests that the LLNL mission can be broadened, and offers advancement of

innovation potential for the ideal research laboratory. The Concept Fusion model

(Cutburth, 1997) offers methods to expand research and improve upon management

methods for that purpose.

The LLNL Mission Paradox

The LLNL mission statement includes research in a variety of categories. But a significant amount is allocated as Nuclear Weapons Stockpile Stewardship. In addition to

this statement, the LLNL is now building the National Ignition Facility (NIF). This $800 million dollar project is designed to provide improved analysis of nuclear weapon materials, and chain reaction phenomena. The facility is designed to produce sub-miniature nuclear explosions. This will also help continue research on nuclear weapons material, for the purpose of safe storage. The NIF experimentation is designed to replace the underground nuclear tests, and insure safe storage.

Congressional hearings that are cited in Chapter 1, Alternative Futures for the Department of Energy , National Laboratories: The Galvin Report, and National Laboratories Need Clearer Missions and Better Management, and detailed in the literature review, show questions about the mission of the LLNL, as well as the DOE. Nevertheless, research at LLNL continues, along with productivity in items which are not nuclear weapons centered. Therefore, if significant funding is for the one purpose, how would it be possible for the LLNL to be producing the long list of non-nuclear weapon devices

The research discovery of this project demonstrates that the LLNL operates in a paradox. In order to produce the high quality of research needed for the named mission, most of anything else that is not the named mission should be put in the mission, and supported with more funding. Furthermore, it would not be possible to meet the named mission without producing the other beneficial products for society.

In the Concept Fusion model (Cutburth, 1997), which was created to explain the actual operations, a high volume knowledge production operation was demonstrated. Therefore, at the core of the mission of the engineering sciences operations, a knowledge production mission was discovered. The high velocity production of knowledge, is adaptable to any form of product, that subsequently can be used to help provide safe storage of nuclear weapons.

The paradox for the nuclear weapon stewardship is somewhat true for several of the other large program areas at LLNL. Each requires development of products which are offset from the original program

Social Impact

The technology assessment shows many useful products that are developed at LLNL, by means of its core knowledge production capability. A consequence is the ability for this laboratory to disseminate knowledge in several forms. Historically, one method of disseminating knowledge referred to as technology transfer at LLNL. At the core of this is the transfer of core knowledge. At this time, great deal of technology transfers by means of funding approaches have been discontinued. Nevertheless, their knowledge base will continue to grow. This should in some way be transferred in larger quantity to the public, through increased focus on idea and product mutations. It is expected, however, that the central barriers are from Congress and from the public being unaware of the true knowledge production mission found at LLNL.

Conclusions and Recommendations

It was demonstrated that standard business models are not sufficiently detailed for adaptation to management analysis for the needs at LLNL. Classical models were examined, and model and new computer software were tested. Although they were shown to be useful, they were not advanced sufficiently, in categories of aggregated stochastic differential computation systems, to emulate the actual LLNL research activities.

The complex interactive research and subsequent knowledge growth potential at LLNL was shown to exhibit computation needs that are proportional to the complex computation methods to examine nuclear physics processes.

It was shown that the complex stochastic differential mathematical game analysis, if substantially broadened, offers potential for experimental analysis of the LLNL operations, when considering a mission that is toward knowledge and innovation production. It is suggested that a research project be undertaken to experiment and articulate improved analysis.

The Concept Fusion model (Cutburth, 1997) demonstrated that the LLNL research mission could be broadened to include what LLNL actually does. Furthermore, their productivity could be expanded, were the research about knowledge production broadened, based on the Concept Fusion model. In addition, a broadening of the management analysis of the projects would improve the quality and diversity of the products that are generated.

It is recommended that funding be increased toward improvement of the current products, so that they are more fully completed and adapted for consumption. It is suggested that the recognition of the knowledge production mission will allow a greater knowledge about the benefits of the products.

It is recommended that a new review be undertaken about the overall definition of technology transfer. It is suggested that failures in that area are from not understanding, and developing the basis knowledge production

It is recommended that a comprehensive research program be undertaken to improve upon the connection and adaptation of functional product analysis, which is founded on a more fully developed mathematical linguistic. That is, the function of a product that is fully understood in the mathematical

linguistic components will be of higher quality, augment greater knowledge production velocity, and subsequently increase velocity and volume of products for the US.

The mission of nuclear weapon stewardship has transformed the mission of the LLNL. It is suggested that the public be informed about the definition of a paradox that exists at LLNL, and its potential for the benefit of society. It would be hoped that this would allow more enthusiastic and full understanding of the mission of LLNL, and bring about a greater production and dissemination of its knowledge base. to be accomplished,

It was discovered that a mission paradox exists at LLNL. For the which includes nuclear weapons stockpile stewardship, almost all science must be explored, which is not nuclear weapon-centered. This demonstrates that the overall productivity for national good can be expanded. It was found that most of the research programs at LLNL are not classified. This allows expansion in those and more categories.

REFERENCES

Note: Government reports, as in Congressional hearings and laboratory reports, are listed following the general book list.

Note: A list of software products used or considered in this paper is shown after the government reports.

Basar, T., & Haurie, A. (1994). Advances in dynamic games and applications. Boston: Birkhauser.

Bateson, R. N. (1993). Introduction to control systems technology. New York: Merill (Macmillan).

Bendy, &Wakefield (1996). See Rayward-Smith, Modem heuristic searches).

Betz, F. (1996). See Gaynor, G. (1996).

Bhattacharya, A. K. A Modular Approach for Probe-Fed and Capacitively Coupled Multilayered Patch Arrays SPIE Proceedings, 45, (2), 193-286.

Birdsall, C. K., & Langdon, B. A. (1985). Plasma physics via computer simulation. New York: McGraw Hill.

Blatt, J. M. & Weisskopf, V. F. (1966). Theoretical nuclear physics. New York: John Wiley.

Bronshtein, I. N., & Semendyayev, K.A. (1979). Handbook of mathematics. New York: Van Nostrand Reinhold.

Bhattacharyya, A.K. A Modular Approach for Probe-Fed and Capacitively Coupled Multilayered Patch Arrays Society of photooptical engineers, 45. (2), 193-286.

Carboni, R. A. (1990). Planning and managing industry-university research collaborations, New York: John Wiley.

Chemical Rubber Company, (1984): CRC Handbook of chemistry and physics, 65th ed. Boca Raton, FL: CRC press..

Churchland, P. S., & Sejnowski, T. J. (1992). The computational brain. Cambridge, MA: MIT Press.

Comstock, G. (1994). Ethics and Scientific Research. Journal of the society of research Administrators, 24. (2).

Considine, M. & Ross, S. (Eds.). (1964). Handbook of applied instrumentation New York, McGraw Hill.

Cook, T. D. & Campbell, D. (1979). Quasi- Experimentation. Boston: Houghton Mufflin.

Cutburth, R. W. (1993). Systems analysis: how to insure the quality of a product. Greeneville, TN Love From the Sea, Publishing Company.

Cutburth, R. W. (1981). The AFSC dynamic intellect model: Greeneville, TN : Love From the Sea, Publishing Company.

Cutburth, R. W. (1996). Kantian relativity for research measure: Greeneville, TN : Love From the Sea, Publishing Company.

Cutburth, R. W. (1995). Social structure theory and elements of knowledge: Greeneville, TN: Love From The Sea, Publishing Company.

Cutburth, R. W. (1997) Concept Fusion: Greeneville, TN: Love From The Sea, Publishing Company.

Dodge, C., W. (1975). Numbers and mathematics; second edition, Orono, Main: Prindle,
Weber and Schmidt, Inc.

Dolfi, D. W., & Hahn, K. (1996). Parallel Optical Link Organization. SPIE proceedings, CR62. (393-404). CR62

Dudgeon, D., & Merseareau, R. (1992). Multidimensional digital signal processing. New York: John Wiley.

Dresher, M.. (1981). The mathematics of games of strategy. New York: Dover.

Eisburg, R. M. , & Resnick, R. (1974). Quantum mechanics of atoms, molecules, solids, nuclei, and particles. New York, Wiley.

Elliott, D. F., & Rao, K. R. (1982). Fast transforms, algorithms, analyses, applications.

New York: Academic Press.

Etzioni, A., (Ed.), (1992). The active society. New York: The Free Press.

Euclid (300 BC). See Gellert & Kustner (1975).

DeBroglie, L. (1900), See Eisburg & Resnick (1974).

Fink, D. C., & Christiansen, D. (Ed.), (1989). Electronics engineer's handbook,

3rd. Ed. New York: McGraw Hill.min.. Princeton: Princeton University Press.
Q-O
Friedman, J. (1987). Planning in the-public

Fusfeld, H. I. & Haklisch, C. S. (Eds.). (1984). University-Industry research

mteractions. Center for Science and Technology Policy: New York. New York

University.

Gaynor, G.H. (Ed.), (1996). Handbook of technology management. New York: McGraw Hill.

Gellert, W., & Kustner, H., (1975). VNR concise encyclopedia of mathematics. New York: Van Nostrand-Reinhold.

Gen, M., & Cheng, R.(1997). Genetic algorithms and engineering design. New York: John Wiley.Gentilman, R. (1996). Piezo-composite smart panel for active surface control Industrial and commercial applications of smart structures technologies. 2721. Society of Photo-optical and Instrumentation Engineers. 234-239.Gilboy, J., A. (1995). Regulatory and Administrative Agency Behavior: Accommodation, Amplification, and Assimilation. Law and Policy,

Gobin, P.F. & Tatibouet, J. (Eds.), (1996) Third International Conference on Intelligent Materials: 2779, (Introduction).

Gray, D. E., (Ed.), (1963). American institute of physics; handbook; New York: McGraw

Gregory, R. L. (Ed.), (1987). The oxford companion to the mind. New York: Oxford.

Hanson, G. H. (1996). Agglomeration, Dispersion, and the Pioneer Firm." Journal of urban economics. (39).

Hanson, T. A. & Day, J.M. (1994). CD-ROM in libraries: management issues. London; Bowker, Saur.

Harashima, F. (1996). Mechatronics-'What it is, Why, and How'?. Joint transactions of the IEEE/ASME on mechatronics, 1. (1).

Heisenberg,W. (1927), See Eisberg & Resnick (1975)

Henry, B. (Ed.) (1991). Forecasting technological innovation. Dordrecht, Germany: Kluwer Academic Press.

Hillier, F. S., and Lieberman, G. J. (1990). Introduction to stochastic models in operations research. New York: McGraw Hill.

Hines, J., Eberlein, B., Richardson, G., Johnson, D., Richmond, B., Melhish, J., and Ho, R. (1996). Vensim PLE and molecules of structure: e-mail address: Jim Hines@

Interserv. Com. : Leap Tec and Lantana Systems Inc. .

Hond, F. D. & Groenewegen, P. (1996) "Environmental Technology Foresight: New Horizons for Technology Management" Technology analysis and strategic management .

Honig, J. C. (Ed.) (1986). Budgeting for sustainability: Operations research society of america:conference. Linthicum, MD: INFORMS.

Houseman, W., (Ed.) (1977). Quantitative analysis for business decisions. Georgetown, Ontario: Richard Irwin Press.

Hughes, D., (1996). STRATAGEM-A methodology and computer-based tool for strategic regeneration. International journal of technology management,

Hunsperger, R. G. (1983). Integrated optics: theory and technology, Heidelberg: Springer.

International conference on optical diagnostics of materials and devices for opto-, micro-, and quantum electronics. SPIE proceedings. Vol. 2648 : Held in Kiev, Ukraine.

Jain, R. K. And Triandis, H. C. (1990). Management of research and development organizations; Managing the Unmanageable. New York: John Wiley.

John, E. (1995). Access To Environmental Information: Limitations Of The Uk Radioactive Substances Registers. Journal of environmental law, 7. (1).

Jones, D. T. (1994). Theoretical approaches to designing novel sequences to fit a given fold. Protein engineering, 6. (4), 460-466.

Jung, Carl, (1964). Man and his symobls. New York: Doubleday.

Kant, I. (1781/1962). Critique of pure reason: New York: Penguin.

Kloeden, P. E., & Platen, E. (1992). Numerical solution of stochastic differential equations. Heidelberg, Germany: Springer-Verlag.

Kosko, B. (1994). Neural networks for signal processing. New York: Prentice Hall.

Kosko, B. (1992). Neural networks and fuzzy systems: A dynamical systems approach To
　　　　machine intelligence. : New York: Prentice Hall.

Lacy, J. A. (1992). Systems engineering management; achieving total quality. New
　　　　York, McGraw Hill.

Lee, S. M. (1987). Management science. New York, Dryden Press.

Long, D. (1995). Federal engvironmental technology transfer process integration to the
　　　　private sector. Oak Ridge, TN: Walden University Ph.D. Dissertation

Maynard, H.B. (1971) Industrial engineering handbook. New York: McGraw Hill.

Marsden, J., & Tromba, A. (1988). Vector calculus. New York: Freeman.

Mandelbrot, B. B. (1977). The fractal geometry of nature. New York: Freeman.
　　　　Matrix management systems. New York: Van Nostrand Reinhold.

Microelectronic processes, sensors, and controls. SPIE proceedings Vol. 2091: 1994.
　　　　Held in Monterey, CA.

Mitchel, A.R., & Wait, R. (1977). The finite element method in partial differential
　　　　equations, New York: John Wiley.

Monck, C.S.P. (Ed.) (1988). Science parks and the growth of high technology firms.
　　　　London: Croom Helm books.

Nossek, J.A. (1996). Designing an Learning with Cellular Neural Networks.
　　　　International journal of circuit theory and applications, 24: (1), 15-24.

Nusse, H. E., and Yorke, J.A. (1994). Dynamic systems. Heidelberg: Springer.

Okano, T., (1996). Intelligent bio-interface: Remote control for hydrophilic-hydrophobic property of the material surfaces by temperature. Third international conference on intelligence materials: Proceedings, 2779. (34-40) Society of Photo-optical and Instrumentation Engineers.

Handbook of optics. Optical Society of America, (1981) New York: McGraw Hill. Proceeding of the american institute of physics on inertial confinement fusion: 1994. AIP.

Physical concepts and materials for novel opto-electronic device applications II. SPIE proceedings Vol. 1985: 1993. Held in Trieste France. Porter, A., L. (Ed.) (1994). Technology Opportunities Analysis: Integrating Technology Monitoring, Forecasting, and Assessment with Strategic Planning Journal of the society of research administrators, 24. (2).

Rayward-Smith, Osman, I.H., Reeves, C.R., & Smith, G.D. (1996). Modem heuristic searches. New York: John Wiley.

Richardson, D. W. (1982). Modem ceramic engineering : New York: Marcel Decker.

Rudin, W. (1982). Principles of mathematical analysis, 3rd ed. New York, McGraw Hill.

Rumbaugh, J., Blaha, M., Premerlani, W., Eddy, F., & Lorensen, W. (1991). Object oriented modeling and design. New York: Prentice Hall.

Russel, B. (1910), Principia mathematica, New York, N.Y. : Van Nostrandt

Russell, B. (1948/1997), Human knowledge, New York, Rutlatge.

Saaty, T. L. (1983/1995). Expert choice. Professional: Decision Support Software. Pittsburgh, PA: Expert Choice.

Schiff, L. (1968). Quantum mechanics. New York: McGraw Hill.

Schrage, L. (1997). Optimization modeling with LINDO. Pacific Grove: Brooks/Cole.

Schaffer and Eshelman (1996). See Rayward-Smith (1996).

Schroedinger, E. (1927), See Schiff (1968).

Sherman, J.D., & Souder, W.E. (1996). See Gaynor,Girard H. (1996)

Sichel, J. L. (1982). Program evaluation guidelines: a research handbook for agency personnel. New York: Human Sciences Press.

Smallenburg, K., Halman, Joop, I. M, & Van Mai, H. H. , (1996). Towards re-use of knowledge in the concept stage of development. Technology management. .

Steensma, Kevin H. (1996). Acquiring Technological Competencies through inter-organizational Collaboration: An organizational Learning

Prospective.Journal of technology management. 12.

Struening, E. & Guttentag, M.. Editors, (1975). Handbook of evaluation research,. Beverly Hills, Ca.: Sage.

Summerton, Jane. (Ed.). (1994). Changing large technical systems. Boulder, Co: Westview Press.

Sychugov, Vladimir A. (1996). Optimization and control of grating coupling to or from silicon-based optical waveguide. Journal of optical engineering 35. (11), 3092-3100

Tager, U. & Von Witzlegben, (Ed.) (1990). Patinnova' 90; strategies for the protection of innovation, proceedings of the first european congress on industrial property rights and innovation commission of the European communities, Luxembourg and IFO-Institute for Economic Research, Munich, Germany.

Taylor, A. & Oats, T., (1996). Technology as Knowledge-Towards a new Perspective on Knowledge Management. International journal of technology managementfl 1). (3/4).

Theil, H., (Ed.). (1965). Operations research and quantitative economics. New York; McGraw Hill.Third international conference on intelligent materials, SPIE proceedings Vol. 2779: Held in Lyon France.

Thompson, A. Jr., & Strickland, A.J. III. (1987). Strategic management. 4th Ed. . Plano TX: Business Publications, Inc.

Toffler, A.. (1990). Power shift. New York: Bantam books.

Tucsnak, M.. (1996). Regularity and Exact Controllability Beam with Piezo-electric Actuator. Journal of control and optimization, 34.(3), 922-930.

Ueberschaer, R. (1996). A Multiscale Approach to the control of Smart Structures. Industrial and commercial applications of smart structures: proceedings, 2721. Society of Photo- optical and Instrumentation Engineers. 94-105.

Van Wyk, R. (1996). See Gaynor, Girard (1996).

Vincent, Thomas L. (1994). See Basar, & Haurie, (1994).

Volkel, R., (Ed.). (1996). Microlens array imaging system for photolithography. Journal of optical engineering, 35. (11), 3323-3330.

Von Bertalanffy, L. (1968). General systems theory; foundations, developments, applications. New York: George-Braziller.

Von Neumann, J., and Morgenstern, O. (1946/1972). Theory of games and economic behavior. Princeton: Princeton University Press.

Weiss, P. A., (Ed.) (1971). Heirarchically organized systems in theory and practice. New York: Hafner.

Williams, J. D., (1954/1982). The complete strategist. New York: Dover.

Wilson, R. J. , & Watkins, J. J. (1990). Graphs: an introductory approach. New York: John Wiley.

Wilson, W. (1996). Operations research . Pacific Grove: Books/Cole. (ITP).

Winspear, G. G. (1968). The Vanderbilt rubber handbook.. Norwalk CT: Vanderbilt Co.

Zemike, F. (1974). Integrated Optic Switch Optical society of America; (pp. 21-24) Topica meeting on Integrated Optics, New Orleans, L

Zey-Ferrell, M., (Ed.) (1979). Reading dimensions of organizations.environment, context,

structure, process and performance. Santa Monica Ca.: Goodyear Publishing Company.

GOVERNMENT REPORTS

Alternative Futures for the Department of Energy , National Laboratories: The Galvin Report, and National Laboratories Need Clearer Missions and Better Management. A Report to the Secretary of Energy. Hearing before the Sub Committee on

Basic Research . March, 1995. ISBN-0-16-047493-0.

Case Study Evaluations GAO/PEMD-10.1.9: General Accounting Office.

Department of Energy: National Laboratories Need Clearer Missions and Better Management: April 12, 1995. : GAO/RCED-95-10: General Accounting Office.Designing Evaluations. GAO/PEMD. 10.1.4: General Accounting Office.

Engineering Annual Summary: 1996. UCRL-ID-127001: Lawrence Livermore National Laboratory. Federal Technology Transfer Policies and our Federal Laboratories:
        Methods for Improving Incentives for the Technology Transfer at Federal

Laboratories: Joint Hearing, Before the Subcommittee on Technology , and the Sub
      Committee on Basic Research, ISBN-0-16-047665-8, June, 1995.
      FY 1996 Comptroller's Report. UCRL-AR-125806: Lawrence Livermore
      National Laboratory.

Government Auditing Standards: GAO-Yellow book: By the Comptroller General
      of the United States. GAO/OCG-94-9: 1994: General Accounting Office.
      How to Measure Performance: Handbook of Techniques and Tools.
      (TRADE); Prepared by the Performance Based Management Special Interest
      Group. Department of Energy DOE. 1995.

Institutional Plan: FY 1996-2001. UCAR-10076-14: Lawrence Livermore National
      Laboratory. Is Today's Science Policy Preparing us for the Future? Hearing
      Before The Committee on Science. January, 1995. ISBN 0-16-046991-0.
      Laws Relating To Federal Procurement, Amended Through Dec.
31,1994,:US

Committee on National Security.
      National Technical Information Services (NTIS): Catalog of Products and
      Services,

US Department of Commerce. The Evaluation Synthesis. GAO/PEMD. 10.1.2: General
      Accouniting Office.The High Performance Computing and Communications

Program. Hearing before the sub committee on Basic Research. October, 1995.
      ISBN-0- 16-052412-1.

Reshaping the Graduate Education of Scientists and Engineers: NAS's Committee on
      Science, Engineering , and Public Policy Report.. Hearing before the Sub